CAROLLE M. D

A Paradigm Shift

FROM TECHNOLOGY-LITERATE PEOPLE TO PEOPLE-LITERATE TECHNOLOGY

A Paradigm Shift
From Technology-Literate People to People-Literate Technology
All Rights Reserved.
Copyright © 2022 Carolle M. Dalley
v2.0

Dalley Publishing

Paperback ISBN: 978-0-578-26309-0
Hardback ISBN: 978-0-578-26310-6

Registration Number: TXu 2-296-597

Cover Photo © 2022 www.gettyimages.com. All rights reserved - used with permission.

PRINTED IN THE UNITED STATES OF AMERICA

Table of Contents

Synopsis

"*A PARADIGM SHIFT*" proposes that humanity is experiencing an overarching transformation which can be seen in trends in several areas of life including technology, psychology, genetic engineering, demography, business operations, project management, biocentrism and reputation management. Together, these trends in diverse disciplines form a widespread paradigm shift that puts Homo Sapiens on the cusp of generating a new way of being human. Although the trends are being discerned by research in separate disciplines, they have commonalities. What is significant among the commonalities is that while humans are using technology to perform tasks that steadily increase in complexity, technology is bringing about a restructure of the way humans think. Here is a sequence of events in the relationship between humans and technology:

- When humans master the ability to perform a task, the knowledge and skill surrounding that task are offloaded to technology.
- Although humans are offloading the tasks to technology, technology does not perform the tasks in the same way that humans do.
- Technology reorganizes the data, redefines the process and digitizes the performance of a task.
- When humans later interact with the digitized task, it brings

about a restructuring of the way that humans think about the task.
- The proliferation of human tasks being offloaded to technology in multiple areas of life indicates an evolving restructure of human thought.
- As technology digitizes more complex tasks in more areas of life, the restructuring of human thought places Homo Sapiens on a path to generating a new version of humanity.

There are two particular trends that undergird the overarching paradigm shift. They are trends in technology and psychology. Technology research and consulting company, Gartner Incorporated, discerns a trend which goes from people becoming literate about technology, to technology becoming literate about people. Psychologist, Wolfgang Giegerich, discerns a trend that goes from consciousness growing by expansion, to consciousness growing by reorganization. These two trends have an underlying influence in several other areas of life. In Gartner's trend, algorithms are acquiring knowledge about people by mining historical data collected from human activities. In Giegerich's trend, the acquisition of knowledge does not just expand consciousness, it also enables a restructuring of consciousness that fosters new outlooks on life. These two trends form a pervasive theme that runs through paradigm shifts in other areas of life. Here are brief descriptions of paradigm shifts in the areas of technology, psychology, genetic engineering, demography, business operations, project management, reputation management and biocentrism.

Gartner's trend in technology is a movement:

From a model of technology-literate people,

To a model of people-literate technology.

Gartner, Incorporated, is a research and consulting firm that offers advice about technological trends to corporations and government agencies. Gartner's technological trend goes from a situation where people become literate about technology by studying various aspects of automation, to a situation where technology is becoming literate about people by the use of Machine Learning algorithms that mine data collected from people.

Wolfgang Giegerich's trend in psychology moves:

From a focus on semantical level psychology where consciousness grows by expansion,

To a focus on syntactical level psychology where consciousness grows by reorganization.

Wolfgang Giegerich is a psychologist who writes books about Analytical Psychology. In Giegerich's trend, human development has been mainly at a semantical level of psychology, where consciousness develops by incremental additions of content, and meanings of content. The trend is moving to a syntactical level of psychology where consciousness develops by a restructuring of content, accompanied by the acquisition of new outlooks.

Jennifer Doudna's trend in genetic engineering is a movement:

From gene therapy which treats genetic diseases after symptoms appear in specific individuals,

To germline enhancement which can change heritable characteristics in future generations.

Jennifer Doudna's Nobel Prize winning work in genetic engineering points to a paradigm shift that is moving from gene therapy to germline enhancement. Gene therapy is about treating genetic disease in specific individuals after they experience symptoms of the disease. Such treatment helps individuals to recover from a genetic disease. Germline enhancement is about changing the DNA of an individual, not only to prevent a genetic disease in that individual, but also to replicate the modified DNA in that individual's future offspring.

David Goodhart's trend in demography is about a cleavage of Western populations into:

"Somewheres" who define their identity by geographic stability and their heritage, and

"Anywheres" who define their identity by geographic mobility and their accomplishments.

David Goodhart is a demographer who sees a trend in which the populations of Western countries are cleaving into two groups of people he characterizes as "Somewheres" and "Anywheres". The "Somewheres" have anchored identities defined in terms of geographic stability, national citizenship and in-person ties to their family and community. The "Anywheres" have portable identities defined in terms of geographic mobility, global citizenship and achievements such as education and career.

Gartner's trend in business operations moves:

From (Somewhere) Operations in physical offices, for local customers, during specific hours,

To Anywhere Operations in digital offices, for global customers, at any time.

Gartner notices a trend in business operations, where the traditional business model is cleaving into two types of operations: the (Somewhere) Operations and the Anywhere Operations. I substitute "Somewhere Operations" for Gartner's traditional "Operations" because Gartner's business model trend resonates with David Goodhart's demographic cleavage of Western populations into "Somewheres" and "Anywheres". In Gartner's trend, Somewhere Operations involves a traditional physical infrastructure for employees to work in a specific geographic location, serving local customers, during specific business hours, while Anywhere Operations involves a digital infrastructure for employees to work from anywhere, serve customers everywhere, and activate operations at any time.

Project Management Institute's trend in project management moves:

From traditional enterprises that have a lesser focus on building a digital culture,

To gymnastic enterprises that have a greater focus on building a digital culture.

The Project Management Institute (PMI) conducted a survey of project managers whose answers reveal a trend in which enterprises fall into two categories: traditional enterprises and gymnastic enterprises. Traditional enterprises are those that put lesser emphasis on building a digital culture, while gymnastic enterprises put greater emphasis on building a digital culture.

Daniel Solove's trend in reputation management moves:

From a social norm where individuals control their own reputations,

To a social norm where algorithms control people's reputations.

An expert in privacy law, Daniel Solove, discerns a paradigm shift in reputation management. The shift is from a custom of personal reputation management to societal reputation management. In the past, individuals managed their reputation by being selective about what personal information they share in the public arena. The trend is toward reputation management by algorithms which gather personal information from Internet sources and assign reputation scores to individuals, without visibility into scoring criteria or accountability for the accuracy of reputation reports.

Robert Lanza's trend in biocentrism is a movement:

From a focus on individuals who are detached observers of reality,

To a focus on collective groups which are co-creators of reality.

A research scientist in biology, Robert Lanza, discerns a paradigm shift in how science regards human interaction with the physical world. Traditionally, science has regarded humans as detached observers of physical reality. More recently, science is coming to recognize collective groups of humans that are active co-creators of reality.

"A PARADIGM SHIFT" argues that the disciplines of technology, psychology, genetic engineering, demography, business operations, project management, biocentrism and reputation management are all experiencing bifurcations which separate the traditional from the contemporary. The bifurcations indicate that all these disciplines are undergoing transformations enabled by technology and supported by psychology. People are offloading to technology the knowledge and skills they acquired about these disciplines. With the offloading, comes the realization that technology does not perform tasks in the manner that people perform tasks. Technology arranges new data storage, defines new processes and digitizes the performance of tasks. When people later interact with the digitized tasks, it engenders a restructure of the way that people think about the tasks. The restructuring of human thought, that accompanies the transference of tasks from people to technology, is reflected in the psychological trend toward a restructuring of consciousness that generates new outlooks on life.

Biological systems, such as Homo Sapiens, are self-organizing systems that evolve through their own internal processes. The order and structure of a self-organizing system develop from within the system, where patterns are formed through interactions among components of the system. Homo Sapiens is a species that developed its order and structure from areas of life that make up human society. Areas of life such as technology, psychology, demography, business operations, genetic engineering, project management, biocentrism and reputation management are among the components of human society. These areas of life are local components of human society, each with its own internal pattern formation. Each component accommodates human activities that are performed according to a shared understanding, or commonly held beliefs, which shape that area of human society. When collective changes in a self-organizing system reach a critical point, they create turmoil in the system and bring about bifurcations, which create the potential for growth. A characteristic of a self-organizing system is that when pattern formations of local components change enough to bring about a disturbance, a new global pattern emerges with greater complexity. "*A PARADIGM SHIFT*" describes changes in the pattern formations of technology, psychology, demography, business operations, genetic engineering, project management, biocentrism and reputation management. This book argues that, taken together, the bifurcations in these components of human society point to Homo Sapiens being on the cusp of generating a more complex version of humanity.

Technology and psychology are like threads running through all the other areas of life in human society. While the relationship between people and technology is changing, there is a corresponding change in psychology. As technology reorganizes knowledge and redefines processes about a particular area of life, human thinking about that area of life becomes restructured. The changes in technology provide support for Wolfgang Giegerich's observation that technological progress is really about an increase in consciousness along with a higher degree of complexity for humanity.

Preface

The theme of this book is that paradigm shifts in various areas of life point to an upcoming emergence of a more complex version of humanity. Homo Sapiens is evolving into a new version of the human species. As humans offload tasks to technology, and technology reorganizes the tasks, humans come to restructure their own thinking about those tasks. Concurrent with the restructuring of human thought are bifurcations in various areas of life which indicate that Homo Sapiens, a self-organizing system, is undergoing change and that a new version of humanity is about to emerge.

The goal of this book is to highlight bifurcations in various areas of life that point to Homo Sapiens being on the cusp of generating a new version of humanity.

The scope of this book is a selection of disciplines where bifurcations are having significant impact on Homo Sapiens. The disciplines are technology, psychology, genetic engineering, demography, business operations, project management, biocentrism and reputation management.

The audience for this book is educated laypeople who have an interest in the relationship between technology and psychology. Readers need not have any specialized knowledge in technology, psychology, or any of the other disciplines. I explain the terms that I use, both in the chapters and in the Glossary.

The arrangement of this book:

> The chapters do not need to be read in the order presented. Where readers are familiar with the bifurcation in any particular discipline, they may wish to skip the material presented in that chapter.

> Chapter 1 describes Gartner's trend in technology.

> Chapter 2 gives an account of Wolfgang Giegerich's trend in psychology.

> Chapter 3 is about Jennifer Doudna's trend in genetic engineering.

> Chapter 4 is an account of David Goodhart's trend in demography.

> Chapter 5 narrates Gartner's trend in business operations.

> Chapter 6 presents the Project Management Institute's trend in project management.

> Chapter 7 depicts Daniel Solove's trend in reputation management.

> Chapter 8 outlines Robert Lanza's trend in biocentrism.

> Chapter 9 is a conclusion based on observations about bifurcations in earlier chapters.

The Cover Design reflects the collective trends in various disciplines that point to Homo Sapiens being on a path to generating a new version of humanity.

Acknowledgements

I am grateful to Michael R. Caplan, a writer who has studied the work of Wolfgang Giegerich for three decades, and who has presented papers at the International Society for Psychology as the Discipline of Interiority (ISPDI) since its inaugural conference in 2012. Michael was generous in offering insightful observations about the nuances of Giegerich's interiorization process as it is applied to the phenomenon of technology.

The UPS Store # 4670 did the graphic design work that made my hand-drawn diagrams into digital images.

An Outskirts Press graphic designer created the cover design for this book.

Outskirts Press staff members were very supportive in the publication of this book. They guided me through the sequence of publishing activities, made recommendations in areas unfamiliar to me, then produced the eBook for available platforms as well as paperback and hardcover formats.

Introduction

While reading about current affairs, I noticed that a corporation and a psychologist are making separate observations about what appears to be similar paradigm shifts. The corporation, Gartner Incorporated, makes its observation about a changing relationship between technology and people. The psychologist, Wolfgang Giegerich, makes his observation about a change in the level of psychology at which people pitch their attention. On further reading about current events in other disciplines, I notice that experts are writing about paradigm shifts in other disciplines. These disciplines include demography, business operations, genetic engineering, project management, biocentrism and reputation management. What makes these disciplines noticeable is that they are experiencing bifurcations, or splits, which are separating traditional outlooks from contemporary outlooks, in diverse areas of life. The contemporary outlooks in these disciplines are built on digital infrastructures that support the bifurcations by changing the relationship between Homo Sapiens and technology. I propose that when taken together, these bifurcations point to Homo Sapiens being on the cusp of generating a new version of humanity.

The paradigm shifts have commonalities which indicate how the relationship between technology and Homo Sapiens is changing. Here are some of the commonalities among the disciplines:

- **Disciplines experiencing bifurcations:** Each discipline is experiencing a bifurcation that splits the discipline into a traditional outlook and a contemporary outlook, which are so disparate they are causing disruptions in various areas of life for Homo Sapiens.
- **Technology increasing the bifurcations:** Homo Sapiens has been offloading to technology an increasing number of tasks that used to be performed by humans. As the volume of offload increases, the effect of the bifurcations also increases.
- **Technology reorganizing tasks:** Technology does not perform the offloaded tasks in the same way that Homo Sapiens performed the tasks. Technology reorganizes the data storage, redefines the processes and digitizes the performance of the tasks.
- **Human thought being restructured:** When Homo Sapiens interacts with the tasks that have been digitized by technology, the impact is that human thought is being restructured.

In writing about these disciplines, I explain the trends discerned by experts, I point out the bifurcations, I provide examples of the tasks being digitized by technology, and I describe the areas of life where human thought is being restructured. I also point out that, when taken together, these disciplines point to Homo Sapiens being on a path to generating a new version of humanity. I devote a chapter of this book to each discipline that I select. Here are my summaries of the paradigm shifts discerned by experts in those disciplines.

This is my summary of Gartner's trend in technology.

There is a paradigm shift …

From a model of Technology-Literate People,
where individuals acquire knowledge about digitalization,

To a model of People-Literate Technology,
where algorithms acquire knowledge about people.

This is my summary of Wolfgang Giegerich's trend in psychology.

There is a paradigm shift …

From a focus on the semantical level of psychology,
where individuals engage in the individuation process,
a goal-seeking effort to differentiate their minds from the
unconsciousness of their communities,

To a focus on the syntactical level of psychology,
where human culture engages in the interiorization process,
an intellectual discipline of interpreting phenomena
that emerge in the world.

This is my summary of David Goodhart's trend in demography.

There is a cleavage of Western populations into …

A population of "Somewheres" who define their identity
by their geographic stability and
their heritage, and

A population of "Anywheres" who define their identity
by their geographic mobility and
their accomplishments.

This is my summary of Gartner's trend in business operations.

There is a paradigm shift …

From a business model of (Somewhere) Operations
where individual store managers enable staff

to work in specific stores,
and serve customers in specific geographic locations,
during specific hours,

To a business model of Anywhere Operations
where computer networks enable staff
to work from anywhere,
and serve customers everywhere,
and activate operations anytime.

This is my summary of Jennifer Doudna's trend in genetic engineering.

There is a paradigm shift …

From a practice of gene therapy that treats genetic
diseases in specific individuals,

To a practice of germline enhancement that changes
heritable characteristics in future generations.

This is my summary of the Project Management Institute's trend in project management.

There is a paradigm shift …

From project management in traditional enterprises
that have a lesser focus on building a digital culture,

To project management in gymnastic enterprises
that have a greater focus on building a digital culture.

This is my summary of Robert Lanza's trend in biocentrism.

There is a paradigm shift ...

From a focus on consciousness as an attribute of
individuals who are detached observers of reality,

To a focus on consciousness as an attribute of collective
groups who are active co-creators of reality.

This is my summary of the Daniel Solove's trend in reputation management.

There is a paradigm shift ...

*From a social norm where individuals
control their own reputations,*

*To a social norm where algorithms
control people's reputations.*

In my opinion, these experts are all observing the same overarching paradigm shift, but from the points of view of their own disciplines. In essence, the paradigm shifts are portrayals of bifurcations in the disciplines of technology, psychology, demography, business operations, genetic engineering, project management, biocentrism and reputation management. These disciplines are diverse. They are not usually related to each other. However, when examined for their commonalities, the paradigm sifts in all these diverse disciplines come together as one overarching paradigm shift. In this book, I explain how I see bifurcations in these disciplines pointing to Homo Sapiens being on the cusp of generating a new version of humanity.

ONE

Gartner's Trend In Technology

There is a paradigm shift ...

From a model of Technology-Literate People,
where individuals acquire knowledge about digitalization,

To a model of People-Literate Technology,
where algorithms acquire knowledge about people.

Gartner, Incorporated, is an organization that conducts research and offers consulting services about trends in technology. One of Gartner's Top 10 Strategic Technology Trends announced in 2019 is stated in terms of a paradigm shift:[1]

There is a paradigm shift ...

From a model of Technology-Literate People,
where individuals acquire knowledge about digitalization,

To a model of People-Literate Technology,
where algorithms acquire knowledge about people.

The paradigm shift is about a change of focus from a model where individuals acquire knowledge about technology, to a model where technology acquires knowledge about people. In the model where people become literate about technology, individuals acquire explicit knowledge about digitalization, and use it to build artificially intelligent systems that are executed by machines. In the model where technology becomes literate about people, algorithms mine large repositories of data to extract explicit and implicit knowledge from historical records of people's behaviors, credentials, subscriptions, online purchases and online searches. In addition, there are algorithms that produce analytics which translate knowledge about people into profiles that are then used to determine their potential as targets for advertisements of consumer goods and for political campaigns. Although the two models of people-being-literate-about-technology and technology-being-literate-about-people continue to co-exist, the acquisition and retention of information is increasingly being performed by technology.

Gartner's Trend In Technology

"Multiexperience" is the name of the technological trend that under-pins Gartner's paradigm shift. Multiexperience technology blurs the boundaries of geography, language and culture to enable collections of people to experience the same reality.[2] Multiexperience is about interaction between people and technology. That interaction is undergoing a significant change due to the use of technology available to capture the multisensory and multimodal experiences of people sharing a common reality. Gartner's observation is that this trend will continue through the 2020's. "The model will shift from one of technology-literate people to one of people-literate technology. The burden of translating intent will move from the user to the computer." said Brian Burke, Vice President of Research at Gartner.[3] He continued "This ability to communicate with users across many human senses will provide a richer environment for delivering nuanced information." The "burden of translating intent" to which Burke refers, will be less about users knowing enough technology to communicate with computer systems, and more about technology using voice recognition, image recognition, and natural language processing, to translate the users' intent. Multiexperience is the technology that is enabling the shift from technology-literate people to people-literate technology.

In my observation, the interaction between technology and people in the late 20th century, was one where people acquired literacy about technology. Now, in the early 21st century, technology is producing algorithms that acquire literacy about people. Multiexperience creates a seamless experience for people by connecting multiple modalities, such as conversational apps, web apps and mobile apps, across a platform of social media channels, web sites, wearables, chatbots, laptops and smartphones, that together, facilitate effortless user involvement. Users are collections of people, for example, employees in a business enterprise, customers of a commercial organization, voters in a democratic election, patients in a health care institution, people being targeted as potential buyers of products and services, students of an educational institution, and citizens of Western countries.

What is noteworthy about Gartner's paradigm shift is that it takes the form of a bifurcation, that is, a split of literacy between human literacy and technological literacy. Literacy used to be attributed just to humans. Now it is also being attributed to technology.

Multiexperience

Gartner's observation is that in the 21st century, there will be a significant change in how users perceive the digital world and how they interact with it via Conversational Platforms, Virtual Reality, Augmented Reality and Mixed Reality.[4] A Conversational Platform is a dialogue management technology that enables a simulated conversation between users and virtual assistants, via chatbot orchestration and natural language processing.[5] In Forbes magazine, contributor Bernard Marr summarizes the three types of reality that underpin Multiexperience.

"Put simply, the difference between virtual, augmented, and mixed reality is this:

- **Virtual reality (VR):** A fully immersive experience where a user leaves the real-world environment behind to enter a fully digital environment via VR headsets.
- **Augmented reality (AR):** An experience where virtual objects are superimposed onto the real-world environment via smartphones, tablets, heads-up displays, or AR glasses.
- **Mixed reality (MR):** A step beyond augmented reality where the virtual objects placed in the real world can be interacted with and respond as if they were real objects." [6]

Jason Wong, Research Vice President at Gartner proposes a four-step model for applying Multiexperience to a digital user journey:[7]

- **"Sync me:** Storing a user's information, which the user can find and access anytime.
- **See me:** Understanding a user's context, location, situation,

historical preferences, and then offering a better information and interaction to the user.

- **Know me:** Using predictive analytics to make suggestions to the user.
- **Be me:** Acting on user's behalf, when given permission, and making the best decision for the user.

The stored information about the users can help business understand their likes and preferences under various contextual situations. This would then enable them to implement predictive analytics for offering recommendations to users for their next course of action."

What I find noteworthy is that in all four steps of the model, "Sync me", "See me", "Know me", and "Be me", the doer is the technology, while the "me" is the human user of the technology. Although the human is the center of attention, it is the technology that is performing all the action. This is not just passive action to store data, but also to conduct predictive analysis and, more assertively, make the best decision for the human. This description of the four-step model focuses on the collection of data about individual users, followed by the use of predictive analytics to offer services based on their preferences. When collected from multiple users, the data can also be used in analytics to target aggregates of users, based on their context, geographic location, situation, and historical preferences. For example, data collected from individuals can be used for marketing hygiene products to gender-specific groups, or for appealing to groups of like-minded voters during an election.

Overall, Gartner's paradigm shift is from individuals becoming literate about technology in the 20th century, to a situation in the 21st century, when a proliferation of Multiexperience enables technology to become literate about people, by capturing data about their histories and their preferences. If new knowledge means knowledge which was unknown to people, then machines can be regarded as bringing forth new knowledge that was implicit in data collected about people's behaviors. To illustrate multiexperience, I will first describe pizza ordering in the technology-literate-people model, then I describe it in the people-literate-technology model.

In the past, people would learn the different ways of ordering pizza and retain the information in their heads, drive into a local pizza shop, or look up the pizza shop in a telephone directory and order on the telephone. To order, they would provide address, name, payment type, pizza type and size. Every time, customers ordered pizza, they would follow the sequence of actions they learned about the technology of pizza production, and provide appropriate information about their order.

In the present, the technology retains information about people for the purpose of ordering pizza through multiexperience development platforms (MXDPs) such as Appian, GeneXus, IBM, Kony, Mendix, Microsoft, Oracle, Outsystems, Pega, Progress, Salesforce, SAP, or ServiceNow. A platform retains information about a customer's order:

- Customer's name, address and telephone number
- Favorite pizza order information, and
- Preferred type of payment.

Domino's AnyWare™

Domino's AnyWare™ is an artificially intelligent system for ordering and delivering pizza via multiple apps and devices. Domino's web site invites customers to order pizza from any of the following devices and apps:[8]

- Google Home – Customers can use Google Home to order by voice.
- Alexa – Customers can use Amazon Alexa to order by voice.
- Slack – Customers can use Slack to order for a team.
- Messenger – Customers can order using FACEBOOK MESSENGER.
- Zero Click – Customers can order using the Zero Click app.
- Text – Customers can order using an emoji that has the image

of a pizza slice.

- Tweet – Customers can order using hash tag #Dominos.
- Car – Customers can order from a FORD car with SYNC and AppLink.
- Smart TV – Customers can order using Domino's app for Samsung Smart TVs.
- Voice Assistant – Customers can order via DOM which is an order-taking expert for ordering via Android or iPhone.
- Smart Watch – Customers can order via APPLE WATCH or ANDROID WEAR.
- Desk Top – Customers can order by going to web site Dominos. com.
- MOBILE & IPAD – Customers can tap REORDER on their personalized homepage.

These devices and apps enable customers to build new orders and re-peat past orders. Customers can order via voice, type, text, click, or tweet. Customers can place orders from home, while watching TV or while driving a car. They can track the orders from Domino's Pizza shops to the delivery locations. The Domino's AnyWare™ technology is becoming literate about people who order pizzas. Multiexperience development platforms enable Domino's AnyWare™ technology to be-come literate about pizza eaters.

Domino's AnyWare™ has a feature named "Pizza Profile" which gives customers the option of receiving e-mail or text messages with recom-mendations based on the customer's pizza profile. In addition, there are messages about available coupons and new pizza-related products coming on the market.[9] There is a "Piece of the Pie Rewards" feature that rewards customers' loyalty with points that can be redeemed for more pizza. The "Easy Order" feature allows customers to repeat their favorite order. A "CAL-O-METER" feature provides customers with calorie counts for their orders. The "Domino's Tracker" feature allows customers to track their orders until delivered to their doors.

Garner's four-step model of "Sync me", "See me", "Know me", and "Be me" is reflected in the assembly of features in Domino's AnyWare™ technology. Domino's AnyWare™ stores information about customers to help the Domino's Pizza company to understand their preferences among several options in the ordering and delivering of pizza. The stored information enables Domino's Pizza to conduct predictive analysis about offering recommendations to customers for their future pizza consumption.

Summary

What I see happening in Gartner's paradigm shift is that literacy is bifurcating, or splitting, into two types of literacy: humans who are literate about technology, and technology that is literate about humans. Literacy used to be an attribute of humans. Now, literacy is splitting between humans and technology. While humans and technology learn in different ways, they are capable of learning about each other.

As a dynamical system, technology is made up of variables whose patterns of behavior change over time. When a variable in a dynamical system bifurcates, it makes the system unstable and therefore predisposed to change.[10] If the magnitude of the bifurcation is great, it can cause a major transformation in the system. Literacy is a variable in technology. The bifurcation of literacy has the potential to bring about a significant change in the relationship between people and technology.

NOTES

1. See Gartner's web site for the paradigm shift from Technology-Literate People to People-Literate Technology: https://www.gartner.com/en/newsroom/press-releases/2019-10-21-gartner-identifies-the-top-10-strategic-technology-trends-for-2020.

2. See Forbes web site for details about Multiexperience technology blurring the boundaries of geography, language and culture to enable collections of people to experience the same reality: https://www.forbes.com/sites/forbestechcouncil/2020/09/21/digital-twin-technology-a-bright-multiexperience-use-case/#7c5b0a956f42.

3. See the quotation from Brian Burke, Gartner's Vice President of Research, on Gartner's web site: https://www.gartner.com/en/newsroom/press-releases/2019-10-21-gartner-identifies-the-top-10-strategic-technology-trends-for-2020.

4. See the Gartner web site for a description of how users will perceive the digital world through Conversational Platforms, Virtual Reality, Augmented Reality and Mixed Reality: https://www.gartner.com/en/newsroom/press-releases/2019-10-21-gartner-identifies-the-top-10-strategic-technology-trends-for-2020.

5. See Gartner web site for comments about Conversational Platform: https://www.gartner.com/reviews/market/conversational-platforms.

6. See Article by Bernard Marr on Forbes magazine web site: https://www.forbes.com/sites/bernardmarr/2019/07/19/the-important-difference-between-virtual-reality-augmented-reality-and-mixed-reality/#1f440c8735d3.

7. See proposal for four-step model on Cigniti Technologies web site: https://www.cigniti.com/blog/multiexperience-enterprise-digital-transformation/.

8. See web site: Domino's ANYWARE, Domino's AnyWare (dominos.com).

9. See web site: Domino's ANYWARE, Domino's AnyWare (dominos.com).

10. See "*Dynamics, Bifurcation, Self-Organization, Chaos, Mind, Conflict, Insensitivity to Initial Conditions, Time, Unification, Diversity, Free Will, and Social Responsibility*" by Frederick David Abraham in "Chaos Theory in Psychology and the Life Sciences", edited by Robin Robertson & Allan Combs, Lawrence Erlbaum Associates, 1995, p 156.

TWO

Wolfgang Giegerich's
Trend In Psychology

There is a paradigm shift ...

From a focus on the semantical level of psychology,
where individuals engage in the individuation process,
a goal-seeking effort to differentiate their minds from the
unconsciousness of their communities,

To a focus on the syntactical level of psychology,
where human culture engages in the interiorization process,
an intellectual discipline of interpreting phenomena
that emerge in the world.

In this chapter, I describe Wolfgang Giegerich's paradigm shift in psychology. Then I use it to give psychological grounding to Gartner's paradigm shift in technology. I also offer my thoughts on the bifurcations occurring in both paradigm shifts, in support of my proposal that concurrent bifurcations in multiple disciplines put Homo Sapiens on the cusp of generating a new version of humanity.

The sources of information that shape my thinking in this chapter are:

- "*Technology and the Soul*" by psychologist Wolfgang Giegerich.[1]
- "*What is Soul?*" by Wolfgang Giegerich.[2]
- "*The Soul Always Thinks*" by Wolfgang Giegerich.[3]
- "*The Tree of Knowledge*" co-authored by biologist Humberto Maturana and cognitive scientist Francisco Varela. [4]
- Michael R. Caplan, a writer who has studied the work of Wolfgang Giegerich for three decades, and who has presented papers at the International Society for Psychology as the Discipline of Interiority (ISPDI) since its inaugural conference in 2012. Michael was generous in offering insightful observations about the nuances of Giegerich's interiorization process as it is applied to the phenomenon of technology.
- "*A Critical Dictionary of Jungian Analysis*" by psychologists Andrew Samuels, Bani Shorter and Fred Plaut.[5]

Wolfgang Giegerich characterizes his paradigm shift as one of movement from the semantical level of psychology to the syntactical level of psychology. He sees the paradigm shift as a movement of attention from individuals to the Western culture as a whole. This movement is a shift of focus from understanding humanity in terms of individual behaviors, feelings and decision-making, to viewing humanity in terms of collective behaviors of human culture, as well as the meanings embodied in collective activities such as the advancement of technology.

Giegerich's Trend In Psychology

Here is how I summarize Giegerich's paradigm shift in psychology:[6]

There is a paradigm shift ...

From a focus on the semantical level of psychology,
where individuals engage in the individuation process,
a goal-seeking effort to differentiate their minds from the
unconsciousness of their communities,

To a focus on the syntactical level of psychology,
where human culture engages in the interiorization process,
an intellectual discipline of interpreting phenomena
that emerge in the world.

Giegerich's paradigm shift applies to general phenomena emerging in the world. In explaining and illustrating Giegerich's paradigm shift, I choose technology as an example of a phenomenon emerging in the world. The paradigm shift in level of psychology enables the realization of a subtler shift, that is, a shift of consciousness, which makes it possible for the psyche to become aware that the site of its agency is shifting from individuals to the collective whole that pervades human culture.

See Appendix A for a description of consciousness.

Giegerich's paradigm shift involves two levels of psychology, and by implication, two levels of consciousness. Giegerich's semantical level is associated with expanding consciousness with knowledge incrementally, while the syntactical level is associated with restructuring consciousness in conjunction with creating a new point of view. In the shift from semantical to syntactical level psychology, there is an accompanying shift of consciousness. The shift in consciousness enables the psyche to become aware that the site of its agency is shifting from individuals to the collective whole that pervades human culture. With this newfound sense of its own agency, the psyche can shift its focus from its scope of action at the individual level to its scope of action at the cultural level. The shift from individual level to cultural level is about consciousness achieving a new type of awareness, and with it, a new scope of action. While these two levels continue their co-existence, an implicit process is becoming explicit to consciousness. At the individual level, technological progress occurs in an unreflective, unconscious manner, where meaning is trapped in the physical nature of technology. It is unreflective because there is little regard for the meaning of technological progress for humanity. It is unconscious because the meaning of technological progress remains implicit. This presents the psyche with a challenge to release the trapped meaning into consciousness and reveal the explicit meaning of technological progress. At the cultural level, the psyche becomes conscious of the trapped meaning as humans reflect on the meaning of the objective process in which they are engaged while advancing technology. Since the two levels of consciousness continue to co-exist, individuals retain their agency and decision-making in their own lives, while cultural consciousness functions in an abstract way in global systems that are technology-based, such as financial markets, political ideologies, health care systems and economics.

I interpret Giegerich's paradigm shift as a bifurcation of human sense of agency. At the semantical level of psychology, the sense of agency dwells in individuals. At the syntactical level, the sense of agency resides in human culture as a whole. The paradigm shift in level of psychology enables the psyche to become aware that the site of its agency is shifting from individuals to the collective whole that pervades human culture. At the semantical level, consciousness is challenging itself to achieve a higher, more differentiated status by interaction with phenomena emerging in the external world, for example, technological advancement. The challenge is for consciousness to think more abstractly, and in doing so, raise itself from the level of the personal, or the semantical, to the level of the impersonal, or the syntactical. Technology is one example of phenomena emerging in the world. The development in consciousness parallels the technological development in the external world.

Giegerich uses a language metaphor for going from the semantical to the syntactical.[7] He compares the psychological shift to a literary shift in focus from words in a sentence, to grammar of the sentence. Words represent the content of consciousness; grammar represents the structure of consciousness. At the semantical level, psychology applies to individual minds. At the syntactical level, the scope of psychology, as a discipline of interpretation, applies to all of human culture.

As a product of the psyche, human culture develops in consciousness through transformations that are effected by gradual abstractions in thought. For example, the emergence of the phenomenon of technology involves abstractions about topics that include Artificial Intelligence, Machine Learning and Deep Learning. These are abstractions about intelligence: What is intelligence? Is intelligence an attribute that belongs to living beings? Can intelligence be ascribed to an inanimate object? Can an inanimate object learn on its own, or must it obtain intelligence from humans? If humans and inanimate objects both possess intelligence, how does human culture decide to allocate intelligent tasks between humans and inanimate objects? In the remainder of this chapter, I address these questions from the perspective of Western culture.

The Western collective mind has been a significant agent of transformation of consciousness in global culture. Here are some of the ways in which Western culture influences global culture:

- When products and services developed in Western countries make their way into other countries, they compete with, and change, the local cultural identity. For example, the Internet and e-mail change communications among people and organizations. Other examples are video games and television game shows which change entertainment from spectator events to events that involve participation.
- In Western cultures, online shopping is replacing visits to brick-and-mortar stores. The replacement is influencing Asian and Latin cultures.

- In the world of finance, the Western culture is changing the nature of transactions. Technology bypasses bank accounts and allows farm workers to send money directly to their families in remote locations via smartphones. This is becoming a global trend.
- There are technological products for translating text in one language to another language, also for teaching people a new language. Western language translation products are becoming available worldwide.
- Western universities are moving from in-person classes to on-line classes that are conducted virtually. Since the classes are online, people in any culture with Internet access can partici-pate. Universities in any culture with Internet access can offer online classes.

Increasingly, the fast-paced advancement of technology that is emerg-ing in the 21st century presents human culture with an opportunity to achieve a higher level of consciousness, by thinking through the mix of elements that make up a phenomenon. For the phenomenon of technol-ogy, these elements include:

- Protection of the privacy of people who use social media
- Ethics of ownership and use of personal data
- Society-sanctioned uses of data captured on video surveillance in road traffic
- Ownership of data collected from public and private sources
- Government regulations about the collection and use of data
- Criteria for content moderation by vendors on whose social media platforms users post their opinions
- Visibility given to criteria for content moderation by owners of social media platforms, and
- Determination of who in society should set criteria for content moderation.

These elements of the emerging phenomenon of technology present complex and often conflicting topics that require abstract considerations for the construction of meanings that are acceptable to society.

In "The Soul Always Thinks" Giegerich distinguishes between semantical and syntactical levels of psychology.[8] The semantical level of psychology refers to ego psychology, or classical Jungian psychology, in which the psychological development of the individual mind is mediated by the ego. At the semantical level of psychology, the development of consciousness is about adding content and adding meanings to expand consciousness. The syntactical level of psychology refers to Giegerich's elaboration of Jungian psychology, or the discipline of interiority, in which the psychological development of global culture has been strongly influenced by the development of the Western collective mind. At the syntactical level of psychology, the development of consciousness is about re-structuring consciousness to accommodate a new outlook on the world, by transforming the old outlook and retaining it in a refined form.

The semantical level and the syntactical level of psychology have been co-existing and will continue to co-exist. The focus of attention used to be on the semantical level of psychology, where the individuation process challenges individuals to differentiate their minds from the unconsciousness of family and community, in other words, to realize their unique potentials as individuals. The focus of attention is shifting to the syntactical level of psychology, where the interiorization process challenges human culture to rise to higher consciousness. The individuation process has a specific goal for individuals to achieve psychological maturity by differentiation of unconsciousness. The interiorization process has no specific goal; it is an intellectual discipline that humans use to interpret phenomena that emerge in the world, for example, emerging technology. Humans interpret phenomena by thinking through the mix of contrary elements that conflict with the conventional consciousness that exists when the phenomena emerge.

See Appendix B for a description of individuation.

In the previous chapter, I described Gartner's paradigm shift from technology-literate people, to people-literate technology. Since Gartner's paradigm shift is about the relationship between technology and psychology, I would like to offer a psychological perspective on Gartner's paradigm shift. There are many psychologies: cognitive, developmental, behavioral, evolutionary, clinical, ecological, forensic, social, educational, biological, organizational, global, among others. To provide a psychological perspective on Gartner's paradigm shift, I chose Analytical Psychology because I see it as offering a multi-faceted view of the psychological attributes of humanity. Cognition. Emotion. Intuition. Sensation. Unconscious influences. Intergenerational influences. The sense-making influence of autopoiesis. The bringing forth of knowledge through enaction, that is, from within. In addition, Analytical Psychology acknowledges both the individual mind and the collective mind. The founder of Analytical Psychology, Carl Jung, proposed the individuation process that pertains to individual human beings, while Giegerich elaborated Analytical Psychology to include the interiorization process, which pertains to human culture raising consciousness to higher levels. Here is Giegerich's observation about technological progress as it relates to psychology:

> "What the great opus of technological progress is really about is the deepening of knowledge about reality, i.e., an increase in consciousness, as well as the factual transformation of human existence in the direction of a higher degree of complexity, differentiation, and logicity." [9]

Jung defined the individuation process by which individual minds develop psychologically. Giegerich developed the interiorization process, as a method of interpreting how human culture uses abstract thinking to raise its consciousness to higher levels. In doing so, Giegerich took Analytical Psychology a significant step beyond where Jung left it. I use Giegerich's work to provide a psychological perspective on the culture of technology-literate people that prevailed in the late 20th century, as well as the culture of people-literate technology in the early 21st century.

In my opinion, Giegerich's depiction of the interiorization process offers psychological support for Gartner's paradigm shift. In short, I think Giegerich is telling us that there is more to Analytical Psychology than individuation. There is something beyond the psychological maturation of individuals. There is a psychological evolution of human culture, in which individuals are embedded. My interpretation of Giegerich's interiorization process is that it provides psychological grounding for Gartner's paradigm shift from the era of technology-literate people, to the era of people-literate technology. Since Giegerich did not include drawings to explain his interiorization process, I have taken the liberty of drawing diagrams, first for my own clarity of mind, and then to help my readers understand the interiorization process as it applies to the phenomenon of technology.

Semantical Level & Syntactical Level

For the purpose of this study, I use the expression "subjective psyche" to refer to the psychological component of an individual person, and the expression "objective psyche" (instead of Giegerich's use of "soul") to refer to the psychological component of collective humankind. The subjective psyche and the objective psyche operate at different levels of psychology. Giegerich uses a metaphor in language to distinguish between the two levels. The subjective psyche operates at the semantical level, while the objective psyche operates at the syntactical level. This is how he describes the distinction in relation to a sentence in a language:

"[T]he syntax of a sentence is not a peer to the semantics (i.e., to the contents or meanings of the words of a sentence). You can reach the level of syntax only by negating the whole level of contents, systematically 'forgetting' it, and rising above it (or descending beneath it) to the level of formal structural relations."[10]

"[T]he level of ... syntax ... invisibly governs and animates our modern world. It is the ... form of our collectively lived life ..." [11]

In other words, the semantical aspect of a sentence is about the content of the sentence, that is, the choice of words and phrases along with the meanings of those words and phrases. Similarly, the semantical level of psychology is about the content of consciousness, such as the addition of new knowledge and the meaning of new knowledge. The syntactical aspect of a sentence is about the grammar, or the rules that govern the arrangement of words and phrases used to compose sentences in a language. Similarly, the syntactical level of consciousness is about the structure of consciousness.

What follows are two Figures that show my interpretation of Giegerich's semantical and syntactical levels of psychology, in the context of the phenomenon of technology.

FIGURE 1: SEMANTICAL LEVEL OF PSYCHOLOGY –
Technology-Literate People

CONSCIOUSNESS IS EXPANDED BY THE
INCREMENTAL ADDITION OF
NEW KNOWLEDGE ABOUT
TECHNOLOGY

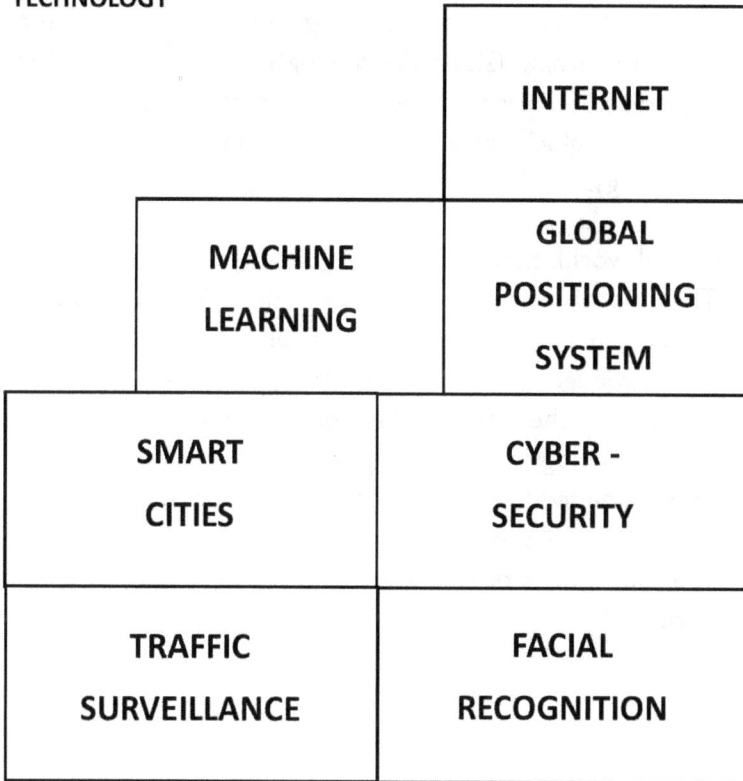

	INTERNET
MACHINE LEARNING	**GLOBAL POSITIONING SYSTEM**
SMART CITIES	**CYBER - SECURITY**
TRAFFIC SURVEILLANCE	**FACIAL RECOGNITION**

CONTEXT: Phenomenon of Technology

SOURCE: Author's interpretation of Wolfgang Giegerich's semantical level of psychology in the context of technology, based on his book "What Is Soul?" [12]

Figure I shows my simple depiction of Giegerich's semantical level of psychology as it relates to the phenomenon of technology. At the semantical level of psychology:

- People are becoming literate about technology, that is, people are the learners and technology is the subject being learned.
- Consciousness is being expanded incrementally by the addition of new topics in technology. To illustrate the incremental addition of content at the semantical level of psychology, I choose topics in the context of technology, for example, the Internet, Machine Learning, Global Positioning System and Cyber-Security.
- The structure of consciousness in Figure I is a building block, with a new block being added when people learn a new topic in technology.

In the external world, there is new technology that is being created by people. The role of learning belongs to people. To create new technology, people acquire literacy about various topics in technology and build their knowledge into artificially intelligent systems. Operating at the semantical level, the psyche integrates new knowledge into the phenomenon of technology by expanding consciousness. At this level, consciousness is expanded by the addition of new topics and new meanings of technology. Figure I shows the expansion of consciousness drawn to look like building blocks of new knowledge being added incrementally, block by block.

FIGURE 2: SYNTACTICAL LEVEL OF PSYCHOLOGY –
People-Literate Technology

CONSCIOUSNESS IS RESTRUCTERED BY
ABSTRACT THOUGHT TO ACCOMMODATE
A NEW OUTLOOK ON
TECHNOLOGY

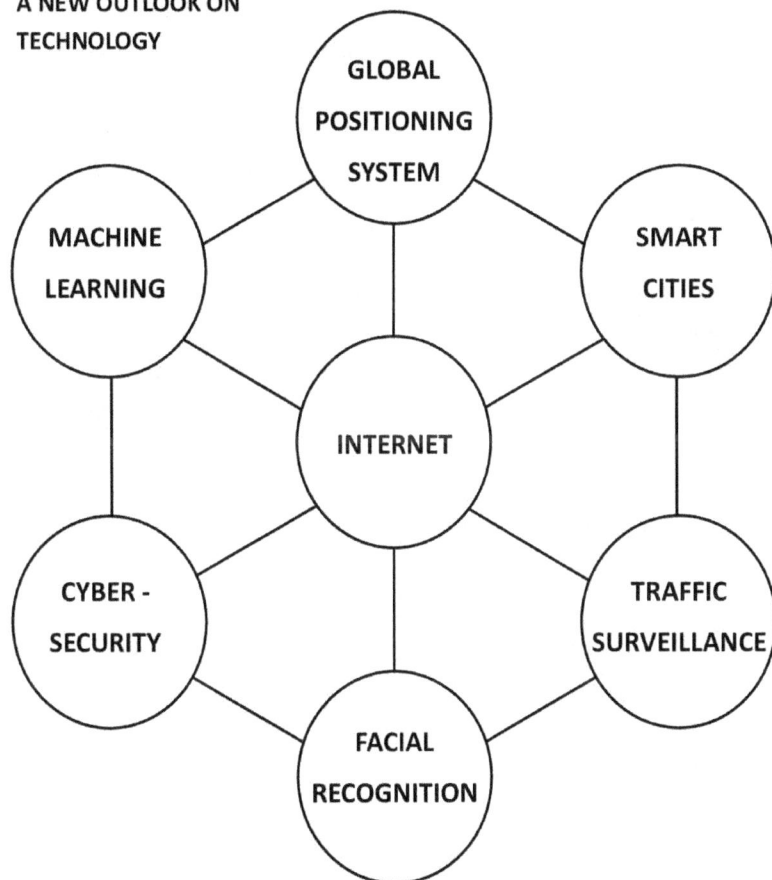

GLOBAL POSITIONING SYSTEM

MACHINE LEARNING

SMART CITIES

INTERNET

CYBER - SECURITY

TRAFFIC SURVEILLANCE

FACIAL RECOGNITION

CONTEXT: Phenomenon of Technology

SOURCE: Author's interpretation of Wolfgang Giegerich's syntactical level of psychology in the context of technology, based on his book "Technology and the Soul".[13]

Figure 2 shows my drawing that depicts Giegerich's syntactical level of psychology as it relates to the phenomenon of technology. At this level of psychology:

- Technology is becoming literate about people, that is, algorithms are the learners and people are the subject being learned.
- Consciousness is being restructured to accommodate a new outlook on the phenomenon of technology. To illustrate the restructure of consciousness at the syntactical level of psychology, I choose a structure that is reflective of technology, a network.
- The building block structure of consciousness depicting the semantical level in Figure 1 is restructured into a network structure depicting the syntactical level in Figure 2.

The role of learning is shifting from people to technology. Technology now includes algorithms that have the capability to learn without help from humans. Moreover, the algorithms are learning about people. Given access to large repositories of data about people, algorithms can acquire literacy about people. The data sources include public records such as information about births, marriages, professional accreditations and voter registrations. Data are also gathered from online activities including Internet searches, online shopping, magazine subscriptions and club memberships. Machine Learning enables algorithms to mine the data collected about people and extract explicit as well as implicit knowledge about people's behaviors, preferences and histories. Since the data about people are continuously being updated, the algorithms can continuously produce up-to-date histories and profiles of people. Operating at the syntactical level, the psyche interiorizes the phenomenon of technology and restructures consciousness to accommodate a new outlook on technology. At this level, consciousness is restructured by the introduction of new rules that apply to technology which is now capable of acquiring knowledge about people. Figure 1 shows consciousness represented by the shape of blocks, while Figure 2 shows the restructure of consciousness drawn in the shape of a network.

Giegerich's paradigm shift from the semantical level to the syntactical level of psychology proceeds as a historical development in which the psyche and the phenomena in the external world evolve in tandem. In this study, the phenomenon under consideration is technology. The particular area of technology being compared with Giegerich's paradigm shift is the trend discerned by Gartner that there is a movement from a model of technology-literate people, to a model of people-literate technology. Later in this chapter, I will use Giegerich's interiorization process to provide psychological grounding for Gartner's paradigm shift.

Giegerich's Definition of Soul (Objective Psyche)

In Analytical Psychology, the psyche is characterized in two ways: the subjective psyche and the objective psyche.[14] This is how Analytical Psychology makes the distinction between subjective psyche and objective psyche:

- The subjective psyche refers to the personal aspect of the psyche and it depends on an individual. It contains experiences that are internal to the individual and are not knowable to anyone else. Giegerich characterizes the subjective psyche as the psyche reductively narrowed down to the ego, or "the empirical man".[15]
- The objective psyche is an expression used to indicate that the psyche has a non-personal existence as a source of knowledge, insight and imagination to humankind; also, that its contents are objective rather than subjective or personal in nature.[16]

The subjective psyche pertains to the personal experience of the individual. The objective psyche is related to the impersonal experience of collective humankind. The subjective psyches of individuals are all part of the objective psyche of humanity.

In referencing Giegerich's interiorization process, I choose the expression objective psyche partly because it is in common usage in Analytical

Psychology, and partly because it has an abstract, felt-but-not-seen quality that makes it compatible with the branch of technology known as Artificial Intelligence. To elaborate on Giegerich's use of the word soul, used interchangeably with psyche, I assembled this definition from descriptors that I found in his book "What Is Soul?".[17] The most essential characteristic of the psyche is its interiority.[18] It is the notion that "… the perceiving and imagining mind cannot survive; it is at once sublated and becomes thinking consciousness … and becomes accessible only to thought." [19] The psyche has no physical being; it expresses itself through an ongoing contradiction in the tension of opposites that characterize the dialectical movement[20] where there is a simultaneity of interiorization and outwardization.[21]

See Appendix C for Giegerich's use of the expressions "soul", "psyche" and "autonomous mind".

In "The Soul Always Thinks" Giegerich describes the characteristics of the soul (objective psyche).[22] Essentially, the soul is consciousness being restructured by the work of abstract thought. It is not a physical entity. It is an ongoing movement, that is intuited, inferred or surmised. It is the effect of the human mind engaged in self-development of the structure of consciousness. In an ongoing movement that is self-reflective, the thinking flows in a circular motion back to itself. The thought is outside of sensory experience and progresses by means of serial contradiction of opposites. Since it is the work of thought, it is accessible to the human mind. Giegerich describes the evolution of the soul (objective psyche) in terms of a number of soul movements.[23] For the purpose of this study, I choose three of Giegerich's soul movements that relate to the phenomenon of technology:

- **Internal contradiction:** This soul movement proceeds on the basis of internal contradictions between identity and difference within the soul.[24] The movement is made up of serial opposites of contradiction. The internal contradiction is called the dialectical movement. This movement has its non-physical existence only in the realm of thought. It is not a natural movement of any physical substance in time. It is the movement of thought itself. The movement is self-enclosed and abstract. Internal contradiction is the dialectical movement of thought that results in the creation of new thought.

- **Procreation:** This soul movement consists of the soul's need to break its own self-containment within its circular movement and open itself to releasing itself from the sphere of thought, and entering into the real world of nature.[25] The pure thought of the soul wants to immerse itself in the sensory world. I see procreation as the productiveness of the soul's movements in the creation of new technological concepts before they are released into the world.

- **Opus of nature conquering nature:** In this type of soul movement, the soul focuses on one particular work, called an opus, and sustains the effort until there is a shift from the current status of consciousness to a higher status of consciousness.[26] This involves a destructive, negating, sublating work imposed upon the current status of consciousness, as if from outside the containment of the soul, with a view to changing consciousness to a new level of truth, that is, a new level of content. The sense of the current consciousness is sublated and refined, and a new consciousness begins to form.

See Appendix D for detailed descriptions of internal contradiction, procreation and opus of nature conquering nature.

Giegerich uses the soul movements of internal contradiction, procreation and opus of nature conquering nature, to underscore his notion that humans are becoming "redundant":

"Man as an individual is nowadays forced to experience his 're-dundancy' and unimportance, his irrelevance, in his logical status as subject he is now more and more being logically dissolved and disintegrated into 'anonymous,' subject-less processes, webs of

relations, and movement per se, for which the keywords globalization, networking, World Wide Web, communication, information
society, international money transactions serve as indications." [27]

While I do see a relationship between technology and people as resulting in a degree of redundancy, I believe Giegerich's idea that humans are
becoming redundant is an extreme view. He lumps all of humanity into
one homogenous group destined for irrelevance.[28] I see the technology-people relationship through the lens of chaos theory.[29] This is how
I apply chaos theory to the technology-people relationship. Evolution
proceeds via periods of equilibrium, periodically interrupted by chaos such as technological advancements of the Industrial and later the
Digital Revolutions. My interpretation is that there are essentially three
possible outcomes from the chaos of a revolution:[30]

1. One possibility is that some people and some systems are over-
   whelmed by the chaos, and they do not survive. This is the
   segment of the population that fits Giegerich's anticipation that
   people are becoming redundant, or irrelevant, due to technological advances.
2. A second possible outcome is that some people and some systems just barely manage to survive the chaos and continue to
   function at the same level at which they functioned before the
   chaos. These people exit the chaos with the same capabilities as
   when they entered the chaos. They learn nothing from the chaos.
   The level of consciousness at which they function is unchanged.
3. The third possibility is that some people and some systems
   emerge from the chaos functioning at a higher level of consciousness and with more capabilities than they had before the
   chaos. This is the segment of the population that I anticipate will
   own the future.

The outcome of the chaos of a revolution does support Giegerich's
anticipation of humans becoming irrelevant, but his anticipation is just
one of the possible outcomes. Giegerich's soul movements of internal

contradiction and procreation both contribute to the dialectical movement of the whole, that is, the opus. The opus involves internal contradiction and procreation in order to move itself forward. The opus drives the evolution of consciousness. The dialectical movement[31] in the psyche results in a history of the development of consciousness.

See Appendix E for details about Giegerich's description of the dialectical movement.

The outcome of the dialectical movement is either an expansion of consciousness or a restructure of consciousness:

- If the dialectical movement occurs in the context of the semantical level of psychology, the outcome is expansion by the addition of new knowledge to consciousness.
- If the dialectical movement occurs in the context of the syntactical level of psychology, the outcome is a restructure of consciousness.

When there is an expansion, or a restructure of consciousness, the effect is a realization that the previous outlook was inadequate. When the inadequacy becomes evident to consciousness, it promotes the evolution of consciousness. I see the shift from semantical level to syntactical level as the psychological equivalent of Gartner's technological paradigm shift. The shift from semantical to syntactical level entails a new outlook. This is Giegerich's explanation of the paradigm shift from a semantical level to a syntactical level of psychology:

"(W)e can note a fundamental paradigm shift. A shift from a 'semantic' to a 'syntactical' level of psychology, ... from 'imaginable substance' to 'abstract form or constitution'. The (semantic level) ... implies a personalistic psychology because it presumed the human person as the foundation or container or subsisting substrate of the life of the soul. The pulsating life that the latter title (syntactical level) is about is no longer tied to the human person as substrate. It is self-sufficient, so that here psychology

> has finally come home to itself and is no longer alienated from itself. The pulsating life ... is the dialectical, self-contradictory movement of union and disunion. Separation and synthesis represent ... not operations performed by the subjective mind, but operations performed by 'the soul' itself." [32]

The expansion of consciousness comes about as a result of the dialectical movement engaging the interior world and the exterior world to bring about small changes in consciousness, incrementally. The restructure of consciousness results in a new outlook, an awareness that matters are not what they seemed to be in the previous stage of consciousness:[33] The interior world does not belong to any individual, it belongs to human culture, as a whole. The exterior world includes everything in nature and everything made by humans. The dialectical movement proceeds unencumbered by any boundaries between interior world and exterior world. Giegerich informs us that there is no border between the interior world and the exterior world.[34] The dialectical movement is the driver of Giegerich's interiorization process ... a process that is not confined by time or space. Western culture plays a significant role in the development of self-consciousness, that is, reflection upon the nature of human culture in general, and the technological transformations that have accompanied and driven this process forward to its current status.

Background for Giegerich's Interiorization Process

Some background about Giegerich's thinking will set the context for his interiorization process. Giegerich sees modern technological civilization as originating from the Enlightenment, which was a counter-movement to Christianity.[35] Giegerich believes that during the Enlightenment era, modern man liberated himself from the Christian religion and attempted to establish a wholly secular world.[36] As part of that world, technological civilization was a deep concern of Western culture.[37] Giegerich explains that Western man had to apply his intelligence to technological development based on the way that his consciousness was informed by scientific pre-conditions in his society at the time. Giegerich points out that the various technologies were not developed in multiple countries at the same time, or for the same purpose. For example, gunpowder was invented earlier in China than in Europe. The Chinese used gunpowder for fireworks in New Year festivals. When the Eastern Europeans developed gunpowder, they used it as an ingredient for cannons and bombs, for the purpose of fighting. The Chinese and the Europeans were informed by the consciousness of their cultures at the time. Technological inventions are possible in multiple countries, but the use, status and function are fundamentally different because the motifs of the culture inform human consciousness and give it a specific orientation.[38]

Western culture is distinguished from other cultures on the basis that two incompatible historical facts might be connected:[39]

- The Christian religion laid claim to absolute truth obtained from divine providence.
- The development of a technological civilization claimed truth derived from empirical facts.

Western culture needed to reconstitute consciousness and characterize truth in a new form. In that space between religious truth and technological truth, psychological truth found a niche. Western culture was experiencing a diminished sense of value in areas that had provided purpose and meaning to people's lives. The satisfaction once derived from myths, rituals and religious practices that had provided grounding in the world, were no longer satisfying. Science was coming into ascendency and it did have value in the sense that it provided empirical explanations for physical events in the world, but science was regarded as cold and unsatisfying with regard to the meaning of life. There was a loss of the sense of being grounded in a world that gave meaning to life.

Psychology came into being in the early 20th century. It was neither a religion nor a science. What psychology offered, was a way of understanding the meaning of one's individual life, in the context of flux between two broad areas: Christian religion that applied to most of the Western world, and science that applied to all of nature. For several decades of the 20th century, Carl Jung's individuation process satisfied a need for meaning. There was analysis for those who wished to achieve a better understanding of themselves and relationships with others. There was counselling for those who wished to put their lives and relationships on more constructive and meaningful paths. There was psychotherapy for the mentally ill. At the end of the 20th century, Giegerich began to publish his elaboration of Analytical Psychology. Jung had placed the focus on the individual mind. Giegerich puts the focus on the life of humankind that is lived collectively. He brings attention to an insight in Jung's psychology and called it "the discipline of interiority" – an intellectual practice that involves shifting from egoic consciousness to psychological consciousness of phenomena.

The following Figure is my way of depicting Giegerich's interiorization process in the context of the phenomenon of technology.

FIGURE 3: INTERIORIZATION PROCESS

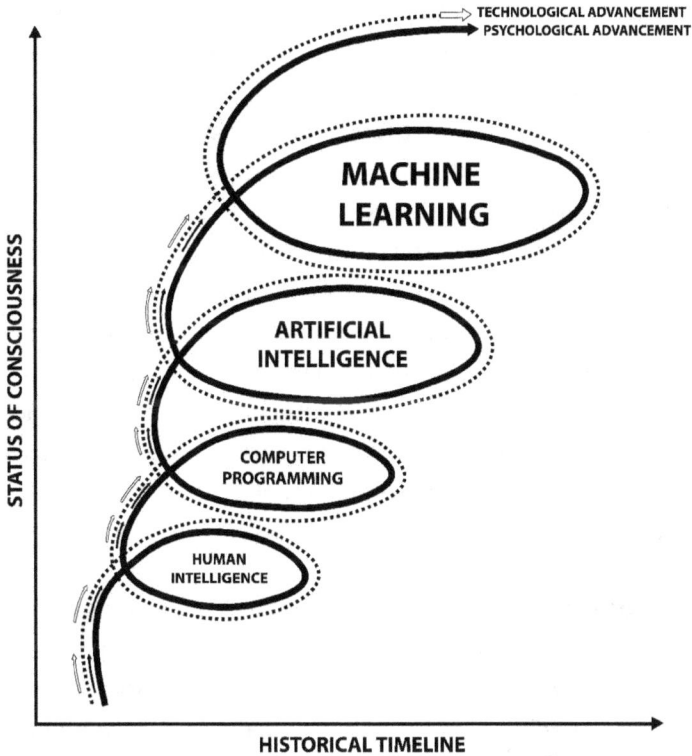

SOURCE: Author's interpretation of the Wolfgang Giegerich's interiorization process as described in "*The Soul Always Thinks*" and applied to the phenomenon of technology.[40]

Figure 3 shows my interpretation of Giegerich's interiorization process applied to the phenomenon of technology. The vertical axis represents status of consciousness. The horizontal axis represents a historical timeline. The spiral shows a solid curve representing the psychological advancement of humankind, while a broken curve represents technological advancement. I draw the two spirals next to each other to indicate my view that psychology and technology advance in tandem. Early in the historical timeline, the first ring of the spiral shows intelligence being regarded as an attribute of humans. This corresponds to the early part of Gartner's technology trend where people are technology-literate. It also represents the early part of Giegerich's psychology trend where ego consciousness prevails at the semantical level of psychology. Later in the timeline, the second ring of the spiral indicates that humans are using their intelligence to develop skills in computer programming. Still later, the third ring of the spiral shows humans developing Artificial Intelligence. Even later, the fourth ring of the spiral shows humans developing Machine Learning, the technology that produces algorithms which are capable of learning about people. This corresponds to the later part of Gartner's technology trend where technology becomes people-literate. It also corresponds to the later part of Giegerich's psychology trend where psychological consciousness prevails at the syntactical level of psychology.

The technology spiral is a reflection of Gartner's prediction of the trend going from a model of technology-literate people to a model of people-literate technology. The psychology spiral is a reflection of the interiorization that the psyche is performing in a dialectical movement over the course of history. Periodically, consciousness recognizes the limitations of its current place in history, thinks through the mix of contrary elements that conflict with conventional consciousness, and moves on to a higher status of consciousness and a new cultural development. Over time, consciousness progressively transforms itself along the history of humanity.

The interiorization process is about focusing attention on what Giegerich calls soul knowing. He states that "… a relentless methodological rejection of egoic knowing, remains indispensable for the constitution of soul knowing." [41] In other words, interiorization is about a shift from ego consciousness to psychological consciousness. Ego consciousness is about the subjective psyche engaging in everyday, practical awareness and interaction with the external world, or having inner thoughts related to the individual. Psychological consciousness is about the objective psyche having a dialogue with itself. In this study, the objective psyche engages in a dialogue through the medium of technology. These two types of consciousness correspond to the semantical and syntactical levels psychology, respectively. Technology is the medium to which I apply the interiorization process.

Giegerich explains the work of ego consciousness and psychological consciousness in terms of *opus parvum* and *opus magnum*:[42] In the next two Figures, I offer drawings to illustrate *opus parvum* and *opus magnum*.

FIGURE 4: OPUS PARVUM

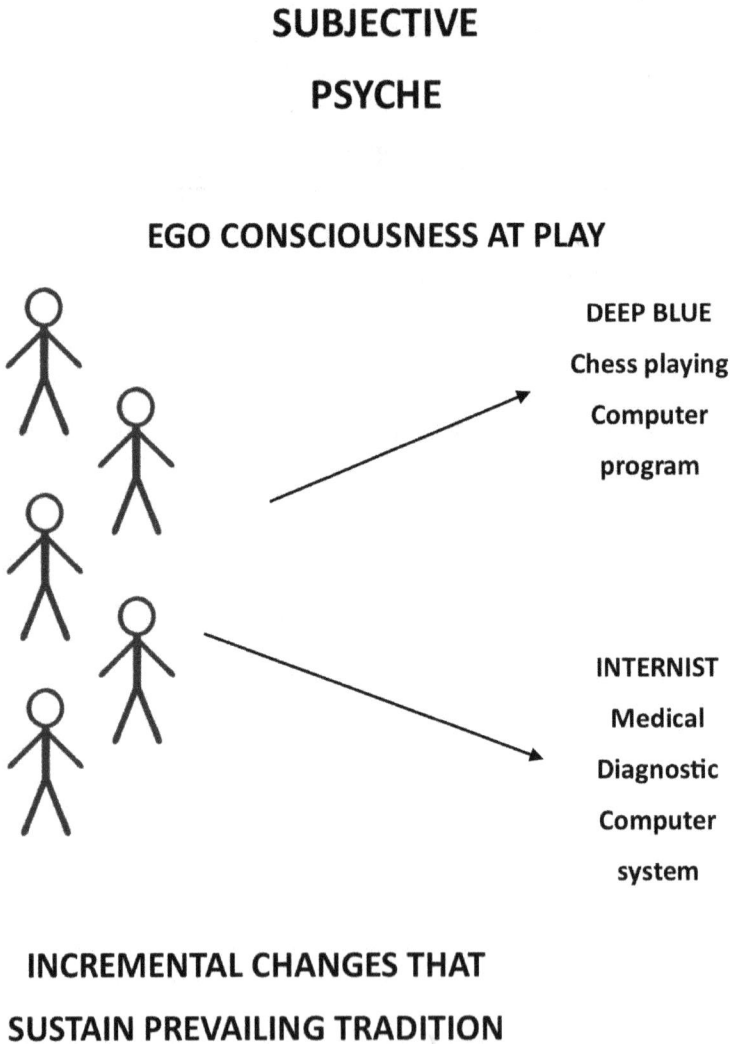

SUBJECTIVE

PSYCHE

EGO CONSCIOUSNESS AT PLAY

DEEP BLUE

Chess playing

Computer

program

INTERNIST

Medical

Diagnostic

Computer

system

INCREMENTAL CHANGES THAT
SUSTAIN PREVAILING TRADITION

SOURCE: Author's interpretation of opus parvum derived from Giegerich's book "*The Soul Always Thinks*".[43]

Figure 4 offers examples of *opus parvum,* which is a small but meaningful work that embodies the accomplishment of an individual, in the context of the individual's life. This type of opus is the work of ego consciousness, and is related to the individual's opinion, motivation and behaviors.[44] This Figure shows the ego consciousness of individuals at play. The stick figures represent the ego consciousness of individuals. At the semantical level of psychology, ego consciousness produces small works that advance technology incrementally. In Figure 4, the subjective psyche shapes the actions of individuals who, having learned elements of technology, develop automated products that mimic the intelligence of individuals. At the semantical level of psychology, ego consciousness produces small works that advance technology incrementally. In this Figure, I show two technological examples of *opus parvum*: DEEP BLUE and INTERNIST. They are the products of ego consciousness. They are the automation of accomplishments of individuals.

- **DEEP BLUE** is artificially intelligent software that automates the game of chess based on rules of the game. IBM hired the 1986 International Grandmaster of chess to articulate his knowledge and skills in playing chess. This led to the development of DEEP BLUE's opening book.[45] The opening book is a database of possible opening moves used in the simulation of human intelligence about the game of chess. DEEP BLUE captured the chess-playing knowledge and skills of the Grandmaster. The ego consciousness of the Grandmaster is at play in producing incremental changes that sustain the prevailing tradition of playing chess. There were other people on the team, but it is the knowledge of the chess Grandmaster that the project sought to capture. DEEP BLUE automates the game of chess, according to the existing rules of the game. DEEP BLUE added knowledge about how the game could be played technologically. One example is the creation of a database of opening moves. Another example is the development of a search engine to select from among possible moves. DEEP BLUE advanced technology incrementally, but did not change the game of chess. What makes DEEP BLUE an example of *opus parvum* is that it increased consciousness by adding knowledge about automated chess playing, but did not result in the restructure of consciousness. Chess continues to be played according to the same rules by which the game had been played traditionally.

- **INTERNIST** is a computerized diagnostic tool designed to capture the knowledge and diagnostic skills that one doctor has about internal medicine, and it was developed at the University of Pittsburg School of Medicine.[46] INTERNIST is an artificially intelligent system that automates the work undertaken by a doctor of internal medicine to collect symptoms about a patient, and compare them with known diseases to recommend a treatment for an illness. INTERNIST mimics the behavior of a human doctor. The ego consciousness of the doctor is at play in producing incremental changes that sustain the prevailing tradition of diagnosing illness. What makes INTERNIST an example of *opus parvum* is that it added knowledge of automated diagnosis to consciousness, but did not have the effect of restructuring consciousness. Internal medical diagnosis continued to be practised as doctors had traditionally conducted diagnoses.

FIGURE 5: OPUS MAGNUM

OBJECTIVE

PSYCHE

PSYCHOLOGICAL CONSCIOUSNESS AT PLAY

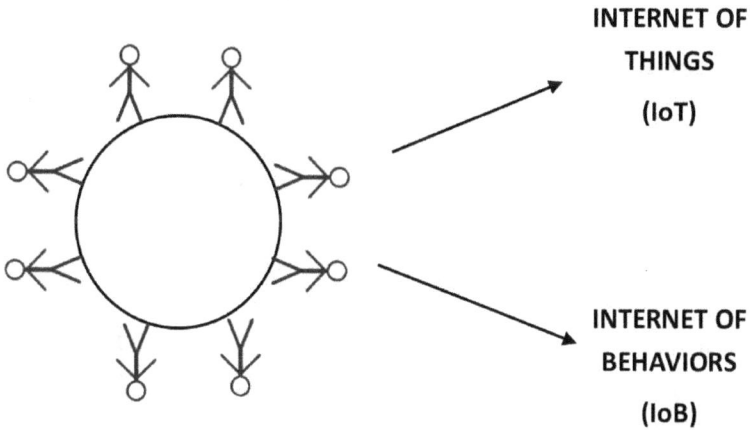

**INTERNET OF
THINGS
(IoT)**

**INTERNET OF
BEHAVIORS
(IoB)**

SIGNIFICANT CHANGES THAT
LEAD TO AWARENESS OF NEED
FOR HISTORIC DEVELOPMENT

SOURCE: Author's interpretation of opus magnum derived from Giegerich's book "The Soul Always Thinks".[47]

Figure 5 shows examples of *opus magnum*. The *opus magnum* is a great work that has significance in a historical context. This type of opus is the work of psychological consciousness.[48] It resonates with people in the generation exposed to it, and possibly later generations. In Figure 5, the objective psyche shapes the actions of human culture, which develops culture-changing products in technology. Psychological consciousness is at play in producing disruptive changes that move human culture to a higher level of sophistication in technology. This Figure shows the outcome of interiorization in the medium of technology. At the syntactical level, psychological consciousness produces great works that have major effects in their historical contexts. This Figure shows two technological examples of great works that impact human culture: The Internet of Things (IoT) and the Internet of Behaviors (IoB).

- **Internet of Things (IoT):** The IoT is a network of physical things which are embedded with sensors, software, and other technologies for the purpose of collecting data and sharing data with other devices and systems through the Internet.[49] The Internet of Behaviors is an extension of the Internet of Things.[50] The IoT gathers data, while the IoB uses collected data to support behaviors.[51]
- **Internet of Behaviors (IoB):** The IoB is a network of technologies that track individuals in terms of biometric data, geographic location and facial recognition, then maps the data to behavioral events via the Internet.[52] The IoB takes data gathered by the IoT and converts it to knowledge about behaviors by mapping IoT data to people's actions.[53] The IoB uses a mix of disciplines: behavioral science, data analytics, and technology. The behavioral science built into the IoB is intended to help people make decisions about topics such as health, physical exercise, relationships, diets and purchases.

An example of the use of the IoT is a health monitor that records data about heart rate and blood sugar levels. An example of the use of the IoB is the production of notifications when the monitor registers data indicating risk to the heart, and offers recommendations for modifying behavior to restore normal heart rate.[54] IoT and IoB are the products of psychological consciousness. They involve algorithms that continuously mine large volumes of data collected about large volumes of people. Algorithms synthesize the data to learn the histories and behaviors of people. This can be done with or without the awareness of the people involved. I characterize IoT and IoB as examples of *opus magnum* because they are of major historical significance. They are great works in the phenomenon of technology and are collectively useful to people across human culture.

Gartner's paradigm shift goes from a model of technology-literate people to a model of people-literate technology. Here is my justification for grounding Gartner's shift in Giegerich's interiorization process:

1. Gartner's model of technology-literate people corresponds to Giegerich's ego consciousness at play, at the semantical level of psychology. The developers of DEEP BLUE and INTERNIST are examples of technology-literate people. DEEP BLUE and INTERNIST are incremental additions to consciousness. The ego consciousness of the developers is at play in the creation of these products.
2. Gartner's model of people-literate technology corresponds to Giegerich's psychological consciousness at play, at the syntactical level of psychology. The Internet of Things and the Internet of Behaviors are examples of people-literate technology. They are acquiring knowledge about people and bringing about a restructure of consciousness. Psychological consciousness is at play as the phenomenon of technology saturates human culture.

The IoT and IoB are built from algorithms that become increasingly literate about people by harvesting data about people, mining the data to extract explicit and implicit knowledge about people, then using the knowledge to support technological services to people.

Giegerich's interpretation of the opus magnum of technology indicates that technology has attained a level of self-reflective consciousness, where it engages in dialogue with itself. Since consciousness attained that level, it means consciousness understands that the elements of technology, regardless of their conflicting natures, are working together toward the self-development of consciousness. The dialogue is in a movement that starts with an initial outlook, and through a series of contradiction of opposites, results in a new outlook. The next Figure is my depiction of the dialectical movement in the interiorization process.

FIGURE 6: THE DIALECTICAL MOVEMENT

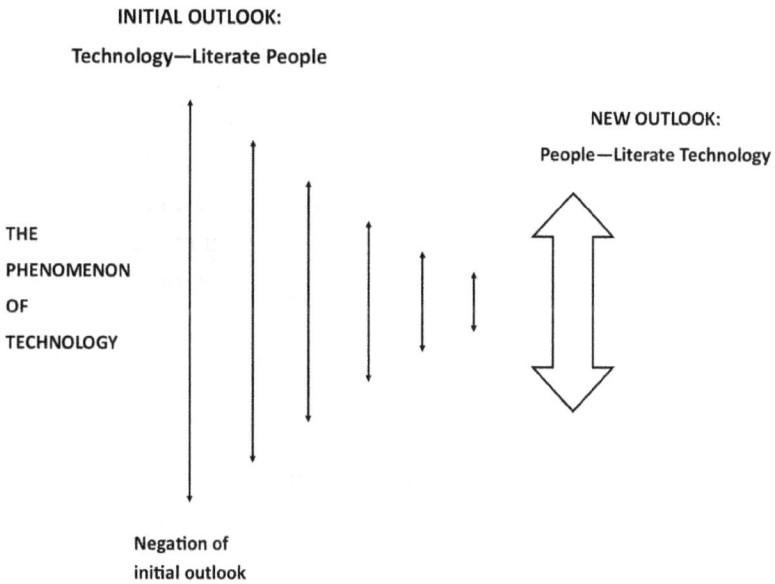

INITIAL OUTLOOK:

Technology—Literate People

NEW OUTLOOK:

People—Literate Technology

THE

PHENOMENON

OF

TECHNOLOGY

Negation of
initial outlook

SOURCE: Author's interpretation of Wolfgang Giegerich's dialectical movement, derived from his book "*What Is Soul?*".[55]

Figure 6 shows my interpretation of Giegerich's dialectical movement[56] applied to the phenomenon of technology. In particular, I apply the dialectical movement to Gartner's paradigm shift from technology-literate people to people-literate technology. The dialectical movement engages in successive cycles of living tension[57] within the objective psyche. The objective psyche engages in a dialogue with itself. Each cycle of the dialectical movement involves the progression from an initial outlook toward a new outlook. In this Figure, the initial outlook is one of technology-literate people. There follows a number of negations of existing outlooks in a series of contradiction of opposites.

In Figure 6, I use double-pointed arrows to show contradiction of opposites. The upper end of each arrow is an existing outlook. The lower end of each arrow is a negation of the existing outlook. The arrows represent the way the dialectical movement goes through cycles of contradictions while progressing from INITIAL OUTLOOK to NEW OUTLOOK. At the end of the cycle of dialectical movement, the new outlook is one of people-literate technology. Each outlook that is negated is retained in a refined form and included in the new outlook. This dialectical movement shows the series of contradiction of opposites that situates Gartner's paradigm shift in a dialectical cycle which:

- Starts with an initial outlook of technology-literate people,
- Undergoes cycles of contradiction in the dialectical movement, and
- Progresses to a new outlook of people-literate technology.

Since Giegerich describes the interiorization process in terms of the concept of autopoiesis,[58] I comment briefly on autopoiesis here. Humberto Maturana and Francisco Varela are Chilean researchers who co-authored a book titled "The Tree of Knowledge" in which they introduced the term "autopoiesis".[59] Maturana and Varela have a view of knowledge acquisition that differs from the traditional meaning. Traditionally, knowledge acquisition is about obtaining knowledge from an external, pre-given world. That knowledge can be decomposed into

items and relationships. According to Maturana and Varela, that definition of knowledge acquisition leaves no room for the creation of new
knowledge by autonomous living beings. Their view of knowledge acquisition is based on the sense-making capability of autonomous living
systems. Varela proposed the term "enactive" to designate this view
of knowledge acquisition to convey the notion that what is known is
brought forth.[60] Giegerich defines the interiorization process in terms
of a sense-making capability that is a feature of autopoietic systems.
He also entertains the notion of knowledge acquisition as a process of
bringing forth, which corresponds to Varela's concept of enaction.

Appendix F has a more detailed description of autopoiesis.

The Confluence of Technology and Psychology

In this section, I describe the confluence of technology and psychology
by pointing out that Giegerich's paradigm shift begins in a context where
there is focus on the subjective psyche, then progresses to a context
where the focus is on the objective psyche. In other words, human culture is progressing from a situation where the individuation process has
been dominant, to one where the interiorization process is coming into
ascendency. Giegerich points to a paradigm shift from a semantical level
psychology to a syntactical level psychology, while Gartner points to a
paradigm shift from a model of technology-literate people to a model of
people-literate technology.

- I see Giegerich's semantical level psychology as being compatible with Gartner's model of technology-literate people. Both
  are focused on the individual, and it is through the individual
  mind that people have been acquiring literacy about technology.
  Semantical level psychology has its basis in the individuation process. The psychology of the individuation process was published
  by Carl Jung in the 20th century. Since then, people have been
  using the individuation process to engage in activities that enable
  them to become literate about topics that are in the attention

of individuals. These topics include writing a computer program, driving a car, interpreting a dream, and practicing active
imagination. I propose that semantical level psychology supports
Gartner's characterization of the 20th century model as being
one of people who are literate about technology.

- I believe Giegerich's syntactical level psychology is compatible
with Gartner's model of people-literate technology. Both are
focused on a cultural reality being shared by aggregates of people. Both are focused on the objective psyche. The syntactical
level of psychology has its basis in the interiorization process.
The psychology of the interiorization process is being published
by Wolfgang Giegerich early in the 21st century. The interiorization process is about the psyche being engaged in the unfolding
of phenomena that are in the attention of human culture. These
phenomena include the unfolding of great technology, great art,
great music, and great architecture. Great technology at this
time in history includes the generation of algorithms that are
capable of acquiring literacy about people, without needing help
from people. I propose that syntactical level psychology supports Gartner's observation of a 21st century trend in which
technology is becoming literate about people.

In bringing together Gartner's technological trend and Giegerich's psychological trend, I invite readers to think about the relationship between
technological progress and human development. Giegerich regards
technological progress as being about an increase in consciousness and
an increase in the complexity of humans. This is how he expresses the
connection between technological progress and human development:[61]

"What the great opus of technological progress is really about
is the deepening of knowledge about reality, i.e., an increase in
consciousness, as well as the factual transformation of human
existence in the direction of a higher degree of complexity, differentiation, and logicity."

The trends observed by Gartner and Giegerich both point to a future about which they articulate an abstract concept. Gartner's abstract concept, people-literate technology, is about an encounter between humanity and technology. Giegerich's abstract concept, syntactical level psychology, is about an encounter between humanity and the environment. Each abstract concept is a reflection of an aggregate of people functioning in a given culture. At the same time, the aggregate of people lives out its history in what appears to be a reflection of the abstract concept. According to Giegerich, the objective psyche is an expression about itself at one moment of ongoing life, through the mouthpiece of the philosopher.[62] Gartner and Giegerich appear to be mouthpieces through which the objective psyche is expressing itself, as it comes into being at this moment in history. This is how Giegerich describes an encounter between humanity and the inanimate environment:[63]

> "Whereas natural life becomes present and fully embodies itself in each living being, each of which is one entity, the life of the (psyche) only comes into being when two fundamentally separate things come together and interact, ...the book and its reader, the dogma and its believer, the music score and the musician or audience, the sounds of a word and the mind that pushes off from them to its meaning. The living (psyche) is in the last analysis what is sparked off when (an inanimate) work and the right human subject touch. The (psyche) is the spark that is ignited. The spark itself is bodiless. It is a happening. Momentary, lasting only for as long as it lasts. It is neither in the (inanimate) work ...nor in the human person, be it in his brain or in the interiority of his mind. No, it is only as the between, as and in their contact, which corresponds to what on the interpersonal level is the sharedness of meanings. It is the encounter between something (inanimate) and the human mind."

Gartner sees the outward expression of the psyche, while Giegerich sees the inward expression of the psyche. This is how Giegerich comments on the simultaneity of the inward and outward expressions:[64]

"In the dialectical movement, there is simultaneity of interiority and outwardization."

As a bundle of continuously changing feelings and representations,[65] the psyche is the encounter between the historical locus and the human mind. That encounter provides an opportunity for the psyche to spark itself, or to give itself a real presence,[66] as it comes into being through self-externalization.[67] Next, I describe the paradigm shift in terms of Giegerich' historical locus.

Historical Locus

In the paradigm shift, the technological trend and the psychological trend both have the same placement in history. Both trends begin in the late 20th century and both indicate a change that progresses during the 21st century. Giegerich uses the expression "historical locus" to refer to one instantiation of the objective psyche at a given time in history. Although Gartner and Giegerich observe the trends from different points of view, I see a parallel between them. In this section, I explain my view that the parallel may be indicative of Gartner's trend linking two instantiations of Giegerich's historical locus:

- Gartner's paradigm shift starts from an era of technology-literate people. I see that as being one instantiation of Giegerich's historical locus, that occurred in the late 20th century.
- Gartner's paradigm shift progresses to an era of people-literate technology. I see that as being another instantiation of Giegerich's historical locus, that is occurring in the early 21st century.

Although we never see our individual minds, we know they exist. Our individual minds are invisible, but we know they exist because we use them every day. We use our individual minds to reflect on our past, perform activities in the present and plan our future. According to Giegerich, our individual minds are all embedded in the psyche. The psyche is also invisible. We know it through its influence because it gives rise to human culture. It gives rise to culture in phenomena such as technology, art, economy, literature and finance. Since it is not a physical entity, the psyche does not exist in a particular geographic location on a particular date in time. I see the psyche as a mental construct, as an abstraction. An abstraction whose existence is not confined to any particular place and time can be viewed in terms of instantiations. Instances of an abstraction can exist in place and time. An abstraction can have multiple instantiations. As an abstraction, the psyche does have instantiations in specific places and times.

While the subjective psyche dominates our stance and our activities in our everyday lives, the objective psyche mediates the historical locus, that is, life lived in the particular culture of a given time and place.[68] "Historical locus" is the expression that Wolfgang Giegerich uses for an instantiation of the objective psyche. The history of the objective psyche is expressed in a series of historical loci that indicate how life is organized and lived in particular cultures at particular points in time.[69]

Giegerich provides a more detailed explanation of the historical locus:[70]

"A given historical locus with its internal unresolved contradictions – for example the contradictions between the actual ideas prevailing in society, on the one hand, and the already changed actual conditions in social life and material reality, on the other hand — demands a progression to new meanings. We may think here of all those situations where the real conditions of life (the actual way of how life is lived) have changed (for example, through technological advances such as through the move from hunting and food-gathering to agriculture, through the move from the handicraft to the industrial mode of production, through the introduction of computers, television, cell phones, GPS, the world wide web), but where the mindset and the categories – the meanings – with which this new situation is perceived are still the ones of the previous, now obsolete cultural status. Real meanings are always necessitated and determined by the actual situation, the new historical locus."

Here I identify two historical loci of the psyche:

- The first historical locus pertains to the culture of the Western world in the late 20th century. This instantiation of the psyche focuses on the phenomenon of technology, in particular, as it pertains to technology-literate people.
- The second historical locus is about the culture of the Western world in the early 21st century. This instantiation of the psyche also focuses on the phenomenon of technology, but particularly on people-literate technology.

In this study, there are two historical loci that are relevant:

- Historical Locus A: The Western outlook, in the late 20th century, I characterize as one instantiation of the objective psyche.
- Historical Locus B: The Western outlook in the early 21st century, I characterize as another instantiation of the objective psyche.

In the next two Figures, I use the image of a cloud to represent each historical locus.

FIGURE 7: HISTORICAL LOCUS A: Technology-Literate People in Late 20th Century

Historical Locus A

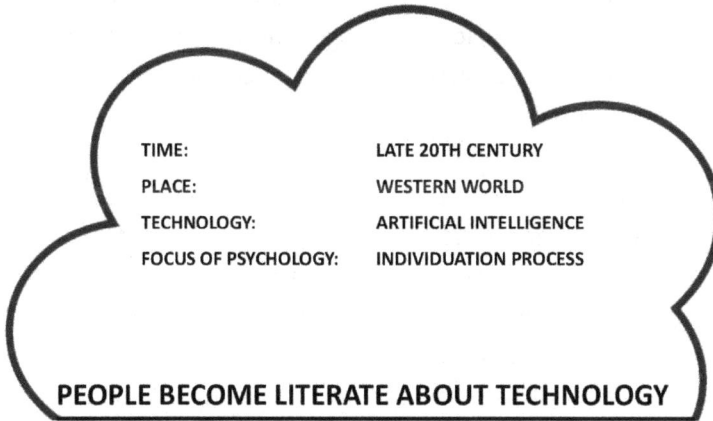

TIME: LATE 20TH CENTURY

PLACE: WESTERN WORLD

TECHNOLOGY: ARTIFICIAL INTELLIGENCE

FOCUS OF PSYCHOLOGY: INDIVIDUATION PROCESS

PEOPLE BECOME LITERATE ABOUT TECHNOLOGY

SOURCE: Author's interpretation of historical locus, based on Wolfgang Giegerich's description in his book "*What Is Soul?*"[71]

Figure 7 shows a cloud representing Historical Locus A in the late 20th century. The characteristics of Historical Locus A are:

- Type of psychology: Ego psychology, or personalistic psychology
- Level of psychology: Semantical level psychology, where the focus is on the importance of the individual, that is, the person, mind and ego
- Temporal dimension: Late 20th century
- Spatial dimension: Western world
- Topics in focus for the individual mind: Artificial Intelligence and ego psychology

Historical Locus A is one instantiation of the psyche, at one point in history, in one geographic area. For the purpose of comparing Gartner's paradigm shift with Giegerich's paradigm shift, I choose to focus on the interaction between two phenomena: technology and psychology. The particular technology is Artificial Intelligence, while the particular psychology is ego psychology. This historical locus is an expression of the psyche when the technological phenomenon of Artificial Intelligence was mimicking the intelligence of individual human beings, for example, an individual playing chess based on the rules of the game. In this historical locus, Giegerich's semantical level of psychology[72] undergirds the expansion of consciousness by its focus on the individual mind. "Semantical" describes the level of psychology where thinking is unique to an individual. This is ego psychology in which the focus is on the mind of the individual. Here are two examples of Artificial Intelligence built to mimic the intelligence of individual human beings in the late 20th century:

- INTERNIST (1979)[73] is a knowledge-based medical diagnostic system that mimics diagnosis as conducted by a human doctor of internal medicine.
- DEEP BLUE (1997)[74] is a computer program that mimics a human chess player well enough to beat the chess world champion.

Historical Locus A occurred in the late 20th century, in Western culture, when the trend was one of people becoming literate about technology. People had to become literate about technology to write computer programs. To write programs, a person needed to have some degree of literacy about the hardware, the operating system, the compiler, the database, and a programing language. Different programing languages have different purposes. Languages that were in common usage in the 20th century were: ASSEMBLER[75] for machine code, FORTRAN[76] for scientific purposes, COBOL[77] and BASIC[78] for business purposes. Different computer programs had to be written for different purposes. A program written for a business purpose was not useful for a scientific purpose. A program written for one business purpose, such as Payroll Processing, was not useful for any other business purpose. There had to be separate programs for Accounts Receivable, Accounts Payable and General Ledger. Similarly, a program written for playing chess was not useful for other games such as GO, Solitaire or Monopoly.

FIGURE 8: HISTORICAL LOCUS B: People-Literate Technology in Early 21st Century

Historical Locus B

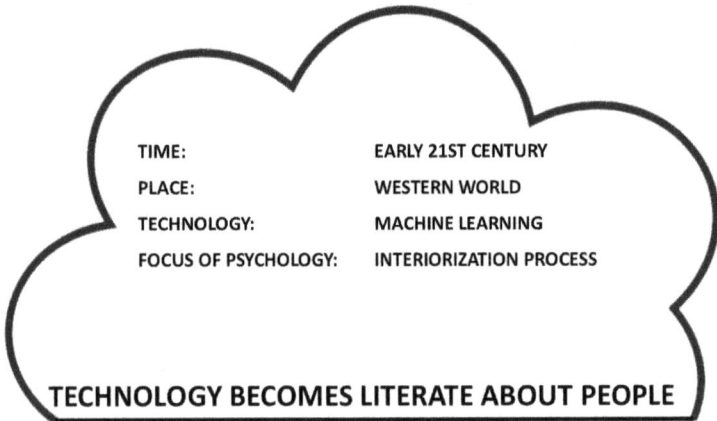

TIME:	EARLY 21ST CENTURY
PLACE:	WESTERN WORLD
TECHNOLOGY:	MACHINE LEARNING
FOCUS OF PSYCHOLOGY:	INTERIORIZATION PROCESS

TECHNOLOGY BECOMES LITERATE ABOUT PEOPLE

SOURCE: Author's interpretation of historical locus, based on Wolfgang Giegerich's description in his book "*What Is Soul?*".[79]

Figure 8 shows a cloud representing Historical Locus B in the early 21st century. These are the characteristics of Historic Locus B:

- Type of psychology: Impersonal psychology
- Level of psychology: Syntactical level of psychology, where the focus is on the importance of human culture, that is, society and the collective mind
- Temporal dimension: Early 21st century
- Spatial dimension: Western world
- Topics in focus for the psyche: Machine Learning and impersonal psychology

Historical Locus B is one instantiation of the psyche, at one point in history, in one geographic area. For the purpose of comparing Gartner's paradigm shift with Giegerich's paradigm shift, I focus on the interaction between two phenomena: technology and psychology. The particular technology is Machine Language, while the particular psychology is impersonal psychology. This historical locus is an expression of the psyche when the technological phenomenon of Machine Learning is generating algorithms that are becoming literate about humans, without the awareness or permission of the people involved. In this historical locus, Giegerich's syntactical level psychology undergirds the restructuring of consciousness by its focus on collective humankind.

"Syntactical" describes the level of psychology which assumes that the psyche has a specific way of thinking that is shared by an aggregate of people or culture, and is not tied to any human person as a physical substrate.[80] This is impersonal psychology in which the focus is on the mind of Western culture. What follows are 21st century examples of the kind of Machine Learning generated algorithms that are capable of acquiring literacy about people. The World Health Organization (WHO) Healthy Cities Project is a global network of cities that employ Internet of Things (IoT) and Internet of Behaviors (IoB) features in an effort to sustain healthy cities. The WHO defines a 'Healthy City' as an urban area that continuously improves the social environments and community resources that enable people to support each other in developing their maximum health potential.[81] They use electronic sensors and smartphones to monitor and collect data from citizens, traffic cameras, city buildings and city assets. The data are analyzed for the purpose of improving the management of community services such as utilities, traffic, hospitals, transportation systems, libraries, schools, waste, power plants, water and sewage, information systems and crime detection. Here are two instances where Gartner's people-literate model is grounded in Giegerich's syntactical level of psychology:

- **IoT-powered Smart Cities (2014):** The WHO Healthy Cities Network has hundreds of member cities around the world harnessing the power of Internet-connected people and things, such as community-led air quality monitoring around the globe using the 'Air Quality Egg' sensor system and crowd-sourced noise pollution monitoring by citizens using a special software running on their smartphones.[82] IoT-powered smart cities aim at improving the quality of life of their populations in a variety of ways, including through measures that promote eco-friendly, sustainable environments and the delivery of health care services to citizens. Wearable wireless health sensors attached to 'smart garments' with specialized conductive areas function, for example, as an axillary body temperature sensor or heart rate/ECG (electrocardiogram) electrodes that wirelessly relay their measurements to remote servers over the Internet. In the second decade of the 21st century, IoT-powered smart cities gathered data. The IoT-powered Smart Cities communicate with citizens about quality of life in Smart Cities across the globe.

- **IoB-powered Smart Cities (2020):** Biometric data are being collected from the WHO Internet-connected Healthy Cities networks of citizens. They use electronic sensors and smartphones to record biometric data gathered from citizens and urban data gathered from city assets. The data serve functions related to public and environmental health surveillance as well as crisis management. There are more than 1,000 Healthy Cities worldwide.[83] In the third decade of the 21st century, IoB-powered smart cities are using gathered data to provide automated services for citizenry, by deploying algorithms that become increasingly literate about people. IoB-powered Smart Cities are capable of monitoring city metrics and sending out notifications to alert citizens worldwide when there are upcoming crises.

Historical Locus B is occurring in the 21st century, in Western culture, when the trend is one of technology becoming literate about people. Machine Learning is enabling algorithms to become literate about people in areas where knowledge of aggregates of people is useful, but the volumes of data are too great for humans to collect and analyze. Algorithms are particularly useful when it is necessary to learn from large volumes of data, without a specific goal and without any idea of what the learning outcome might be. Algorithms extract implicit knowledge from data collected about humans.

Here are more examples of the areas of life where algorithms are becoming people-literate:

- Traffic surveillance – Algorithms collect data from cameras mounted on streets and make decisions about issuing traffic tickets to people who violate the road code.
- Virtual personal assistants – Algorithms collect data about people in order to customize answers to questions that people pose to the automated assistants.
- Recommended purchases – Algorithms analyze data collected from past purchases and use it to make recommendations about future purchases of products such as books and movies.
- Search engines – Algorithms are used to rank the web sites displayed as the result of a search.
- News feeds - Algorithms perform analytics on Internet usage and decide on content to present in news feeds, depending on

past Internet activity.

- Online dating – Algorithms match couples by analyzing their preferences, their descriptions of themselves and their common interests.

In order to perform the services listed above, the algorithms keep track of people's behavior such as products purchased online, words searched, Internet browsing, eTrading history and advertisements viewed. Increasingly, humans are being left out of the decisions because algorithms are becoming people-literate.

Summary

In Wolfgang Giegerich's paradigm shift, I see human agency bifurcating, or splitting, into two types: one type of agency resides in individuals, while the other type of agency resides in human culture. Humans used to understand the psyche as having its site of agency in individual activities. Now, the psyche is becoming aware that the site of its agency is in the collective whole that pervades human culture. Humanity used to be understood mainly in terms of individual behaviors, feeling and decision-making. Humanity is now being understood in terms of collective behaviors of human culture, as well as the meanings embodied in collective activities such as the advancement of technology.

Psychology is a dynamical system that has variables whose patterns of behavior change in the passage of time. When a variable in a dynamical system bifurcates, it renders the system unstable, so the system becomes predisposed to change.[84] If the magnitude of the bifurcation is large, it can cause a significant transformation in the system. Human agency is a variable in psychology. The bifurcation of human agency creates the possibility of a major change in psychology.

Giegerich's paradigm shift in psychology involves a development in consciousness. Consciousness is challenging itself to achieve a higher, more differentiated status by interaction with phenomena emerging in the external world, for example, technological advancement. The challenge is for consciousness to think more abstractly, and in doing so, raise itself from the level of the personal, or the semantical, to the level of the impersonal, or the syntactical. My interpretation is that the bifurcation of the human sense of agency is reflected in Giegerich's observation about the great opus of technological progress being a transformation of human existence in the direction of a higher degree of differentiation and complexity. In "Technology and the Soul" Giegerich makes the observation:[85]

> "What the great opus of technological progress is really about is the deepening of knowledge about reality, i.e., an increase in consciousness, as well as the factual transformation of human existence in the direction of a higher degree of complexity, differentiation and logicity."

I believe that history provides support for Giegerich's observation that technological progress is really about an increase in consciousness along with a higher degree of complexity for humanity. In my reading of history, humanity offloads increasing capabilities of intelligence to technology, from one revolution to the next. From the Agricultural to the Scientific, to the Industrial, and now to the Digital Revolution. Each offloading has freed us from the labor of some skill we have mastered and left us free to pursue novel capabilities that enable us to tackle the next revolution. Now that technology is capable of learning, there seems to be an upcoming opportunity for humanity to move on to the greater complexity and differentiation of which Gingrich writes.

My reading of Gartner's paradigm shift in technology and Giegerich's paradigm shift in psychology is that both point to Homo Sapiens being on the cusp of generating a new version of humanity.

NOTES:

1. See "*Technology and the Soul: From the Nuclear Bomb to the World Wide Web*", The Collected English Papers of Wolfgang Giegerich, Volume 2, Routledge, 2007.
2. See "*What is Soul?*" by Wolfgang Giegerich, Spring Journal Books, 2012.
3. See "*The Soul Always Thinks*", The Collected English Papers of Wolfgang Giegerich, Volume 4, Routledge, 2020.
4. See "*The Tree of Knowledge*" co-authored by Humberto Maturana and Francisco Varela, Shambala 1987.
5. See "*A Critical Dictionary of Jungian Analysis*" by psychologists Andrew Samuels, Bani Shorter and Fred Plaut, Routledge & Kegan Paul Ltd, 1993.
6. See Giegerich's observation of a paradigm shift in psychology in "*The Soul Always Thinks*" pp 329 – 330.
7. See "*What Is Soul?*" p 105 for sentence being used as a metaphor for semantical and syntactical levels of psychology.
8. See Giegerich's distinction between semantical and syntactical levels of psychology in "*The Soul Always Thinks*" pp 329 – 330.
9. See Giegerich's comment about the great opus of technological progress in "*Technology and Soul*" pp 310 – 311.
10. See "*What Is Soul?*" p 105 for sentence being used as a metaphor for semantical and syntactical levels of psychology.
11. See "*What Is Soul?*" p 149 for reference to syntactical level of psychology animating modern world.
12. See "*What is Soul?*" by Wolfgang Giegerich, Spring Journal Books, 2012.
13. See "*Technology and the Soul: From the Nuclear Bomb to the World Wide Web*", The Collected English Papers of Wolfgang Giegerich, Volume 2, Routledge, 2007.
14. See Jung, Collected Works Volume 7, "*The Structure of the Unconscious*", paragraphs 437 – 507 (pp 263 – 292).
15. See "*The Soul Always Thinks*" p 581 for reference to "empirical man".
16. See "*A Critical Dictionary of Jungian Analysis*" by Andrew Samuels et al, pp 100 – 101 for description of objective psyche.

17. See "*What Is Soul?*" pp 6 – 60 for descriptions of "psyche" by Giegerich.
18. See "*What Is Soul?*" p 59 for comment about interiority of the psyche by Giegerich.
19. See "*The Soul Always Thinks*" p 147 for comment about the psyche becoming thinking consciousness.
20. See "What Is *Soul?*" by Wolfgang Giegerich, p 59 for comment about the psyche in relation to the dialectical movement.
21. See "*What Is Soul?*" by Wolfgang Giegerich, p 60 for comment about simultaneity of interiorization and outwardization.
22. See "*The Soul Always Thinks*" Volume 4, by Wolfgang Giegerich, p 308 for descriptions of the soul.
23. See "*The Soul Always Thinks*" Volume 4, by Wolfgang Giegerich, pp 310 – 323 for reference to soul movements.
24. See "*The Soul Always Thinks*" pp 308 – 311 for reference to internal contradiction.
25. See "*The Soul Always Thinks*" pp 311 – 314 for reference to procreation.
26. See "*The Soul Always Thinks*" pp 316 - 323 for reference to opus of nature conquering nature.
27. See "*The Soul Always Thinks*" p 322 for reference to soul movements.
28. See "*The Soul Always Thinks*" p 322 for reference to irrelevance of humans.
29. See "*CHAOS*" by James Gleick pp 243 – 272 for reference to chaos theory.
30. See "*CHAOS*" by James Gleick pp 243 – 272 for reference to chaos theory.
31. To describe the dialectical movement, Giegerich uses the expressions "the dead letter", "the push-off" and "the corpse" all of which imply the existence of a physical substrate. I avoid those expressions because they are contradictory to Giegerich's definition of soul, aka psyche. In my description of the dialectical movement, I choose the expression objective psyche because it does not imply the existence of a physical substrate.

32. See "The Soul Always Thinks" pp 329 – 330 for reference to the soul.
33. See "What Is Soul?" p 105 for reference to consciousness.
34. See "What Is Soul?" p 298 for Giegerich's comment about there being no border between the interior world and the exterior world.
35. See "Technology and Soul" p 159 for Giegerich's comments about modern technological civilization.
36. See "Technology and Soul" p 159 for Giegerich's comment about establishment of the secular world.
37. See "Technology and Soul" p 160 for Giegerich's comment about Western culture.
38. See "Technology and Soul" p 160 for Giegerich's comment about technological inventions in multiple countries.
39. See "Technology and Soul" p 161 for Giegerich's comment about Western culture.
40. See "The Soul Always Thinks" pp 327 – 328 for Giegerich's comment about interiorization.
41. See "What Is Soul?" p 299 for reference to soul knowing.
42. See "The Soul Always Thinks" p 338 for reference to opus parvum and opus magnum.
43. See "The Soul Always Thinks" p 338 for Giegerich's comment about opus parvum.
44. See "The Soul Always Thinks" p 338 for reference to opus parvum.
45. See Big Blue web site IBM100 - Deep Blue.
46. See ACM Digital Library web site Timeline of AI Achievements for information about INTERNIST.
47. See "The Soul Always Thinks" p 338 for Giegerich's comment about opus magnum.
48. See "The Soul Always Thinks" p 338 for reference to psychological consciousness.
49. See Oracle web site: What Is the Internet of Things (IoT)? (oracle.com).
50. See TechTheDay web site: Internet of Behaviors (techtheday.com).
51. See TechTheDay web site: Internet of Behaviors (techtheday.com).
52. See TechTheDay web site: Internet of Behaviors (techtheday.com).

53. See TechTheDay web site: <u>Internet of Behaviors (techtheday.com)</u>.

54. See TechTheDay web site: <u>Internet of Behaviors (techtheday.com)</u>.

55. See *"What Is Soul?"* pp 298 – 305 for Giegerich's comment about dialectical movement.

56. See *"The Soul Always Thinks"* pp 327 – 329 for reference to the dialectical movement.

57. See *"The Soul Always Thinks"* p 328 for reference to living tension.

58. See *"What Is Soul?"* by Wolfgang Giegerich, p 60 for reference to autopoiesis.

59. See "The Tree of Knowledge" co-authored by Humberto Maturana and Francisco Varela, Shambala 1987.

60. See *"The Tree of Knowledge"* pp 253 - 255 for explanation of "enactive" by Varela.

61. See *"Technology and Soul"* pp 310 – 311 for Giegerich's comment on the great opus of technological progress.

62. See *"What Is Soul?"* p 10 for Giegerich's comment about the psyche being an expression about itself at one moment of ongoing life, through the mouthpiece of the philosopher.

63. See *"What Is Soul?"* pp 56 – 57 for Giegerich's explanation of when the psyche comes into being.

64. See *"What Is Soul?"* p 60 for Giegerich's comment about the simultaneity of interiority and outwardization.

65. See *"What Is Soul?"* p 8 for Giegerich's comment about the psyche being a bundle of continuously changing feelings and representations.

66. See *"What Is Soul?"* pp 58 – 59 for Giegerich's explanation of the spark.

67. See *"What Is Soul?"* p 59 for Giegerich's comment about self-externalization of the psyche.

68. See *"What Is Soul?"* p 298 for Giegerich's remark that the psyche dominates the historical locus.

69. See *"What Is Soul?"* p 43 for Giegerich's comment about the psyche being grounded in how life is organized and lived in particular cultures at particular points in time.

70. See "*What Is Soul?*" pp 42 - 43 for Giegerich's explanation of historical locus.
71. See "*What Is Soul?*" p 188 for Giegerich's explanation of historical locus.
72. See "*The Soul Always Thinks*" p 329 for Giegerich's paradigm shift from semantical level psychology to syntactical level psychology.
73. See the ACP Internist web site: http://blog.acpinternist. org/2010/01/artificial-intelligence-for-real.html.
74. See announcement about Deep Blue on the History web site: https://www.history.com/this-day-in-history/ deep-blue-defeats-garry-kasparov-in-chess-match.
75. See description of the ASSEMBLER language on the IBM web site: https://www.ibm.com/support/knowledgecenter/en/SSLTBW_2.1.0/ com.ibm.zos.v2r1.asma400/asmr102112.htm.
76. See description of the FORTRAN language on the FORTRAN web site: https://fortran-lang.org/.
77. See description of COBOL on the Search IT Operations web site: https://searchitoperations.techtarget.com/definition/ COBOL-Common-Business-Oriented-Language.
78. See description of BASIC on The History of the BASIC programming language web site: https://www.thoughtco.com/history-basic- programming-language-1991662#:~:text=%20The%20History%20 of%20the%20BASIC%20Programming%20Language,the%20 mid-1980s%2C%20the%20mania%20for%20programming...%20 More%20.
79. See "*What Is Soul?*" p 298 – 300 for Giegerich's description of historical locus.
80. See "*The Soul Always Thinks*" p 330 for Giegerich's comment that the soul is no longer tied to the human person as substrate.
81. See the International Journal of Health Geographics web site: On the Internet of Things, smart cities and the WHO Healthy Cities | International Journal of Health Geographics | Full Text (biomed- central.com).
82. See the World Health Organization (WHO) web site: WHO | Types of Healthy Settings.

83. See the World Health Organization (WHO) web site: <u>WHO |</u> <u>Types of Healthy Settings.</u>

84. See "*Dynamics, Bifurcation, Self-Organization, Chaos, Mind, Conflict, Insensitivity to Initial Conditions, Time, Unification, Diversity, Free Will, and Social Responsibility*" by Frederick David Abraham in "*Chaos Theory in Psychology and the Life Sciences*", edited by Robin Robertson & Allan Combs, Lawrence Erlbaum Associates, 1995, p 156.

85. See "*Technology and Soul*" pp 310- 311 for Giegerich's comment on the great opus of technological progress.

THREE

David Goodhart's
Trend In Demography

There is a cleavage of Western populations into ...

A population of "Somewheres" who define their identity
by their geographic stability
and their heritage, and

A population of "Anywheres" who define their identity
by their geographic mobility
and their accomplishments.

This chapter is about a paradigm shift in which Western populations are cleaving into two groups that demographer David Goodhart calls "Somewheres" and "Anywheres". The two groups physically resemble Homo Sapiens, but I believe the groups are cognitively and psychologically different from each other. What I find salient about this paradigm shift is that the cleavage of Western populations results in two groups that define their identities in different ways. This demographic paradigm shift is about a bifurcation in the identity of Western populations.

The sources of information that shape my thinking in this chapter are:

- "The Road to Somewhere: The Populist Revolt and the Future of Politics" by demographer David Goodhart[1]
- "Populists in Power Around the World" from the web site of the Tony Blair Institute for Global Change[2]
- "The Meritocracy Trap" by law professor Daniel Markovits[3]
- "The Individuation Principle" by psychologist Murray Stein[4]
- "Sapiens: A Brief History of Humankind" by historian Yuri Noah Harari.[5]

David Goodhart is head of the Demography, Immigration and Integration Unit at the think tank Policy Exchange in the United Kingdom.[6] In "The Road To Somewhere" he assembled his ideas about the cleavage of Western populations based on surveys, statistics about manufacturing jobs, and opinion polls.[7] He began writing the book before the June 2016 BREXIT election in the United Kingdom (UK), and finished it after the November 2016 election in the United States of America (USA).[8] Goodhart informs us that since the last generation, Western society has been cleaving into two groups: the "Somewheres" and the "Anywheres".[9] Here is how I summarize Goodhart's trend in demography:

There is a cleavage of Western populations into ...

A population of "Somewheres" who define their identity
by their geographic stability
and their heritage, and

A population of "Anywheres" who define their identity
by their geographic mobility
and their accomplishments.

Goodhart published his book in 2017, the year after populist leaders won 2016 elections in the United Kingdom (UK) and the United States of America (USA). I am writing in 2021, that is, after the UK BREXIT Deal was agreed in December 2020 and after the assault on the USA Capitol occurred in January 2021.

"Somewheres" and "Anywheres"

Goodhart points out that while we all have a mix of the identities of these two groups, there are characteristics that make the Somewheres distinct from the Anywheres. Based on his description, I see the Somewheres as traditional populations, that have been with us for many generations, while the Anywheres are emerging from recent generations. According to Goodhart, the Somewheres are people who have ascribed identities, while the Anywheres have achieved identities, and the Inbetweens have a mix of both identities.

These are the factors that make up the ascribed identities of the Somewheres:[10]

- They have stable identities in terms of geographic location where they live and by in-person ties to their community.
- They are usually not university educated.
- They are mostly from the working class.

- They feel left behind and experience nostalgia for the past.
- They are uncomfortable with changes brought about by globalization and technology.
- They feel marginalized.
- Geographically rooted, they believe that their needs are not being addressed in their geographic location.
- They tend to be Conservative voters.
- They are uncomfortable with immigration.
- They are uncomfortable with rapid social change.
- They see themselves as citizens of a nation.
- They tend to be socially connected by in-person networks.

These are the factors that make up the achieved identities of the Anywheres:[11]

- They have portable identities in terms of their geographic mobility and achievements such as education and career.
- They are usually university educated.
- They are mostly from the middle class.
- They do not feel left-behind, nor experience nostalgia for the past.
- They are comfortable with changes brought about by globalization and technology.
- They do not feel marginalized.
- Geographically mobile, they go to those geographic locations where they expect their needs can be met.
- They tend to be Liberal voters.
- They are comfortable with immigration.
- They are comfortable with rapid social change.
- They see themselves as citizens of the world.
- They tend to be socially connected by technological networks.

What the Somewheres value about themselves are their ascribed identities, which are stable. They value living in a geographic location in close proximity to where they grew up, and they value their ties to family and community. What the Anywheres value about themselves are their achieved identities, which are portable. They value their geographic mobility as well as their achievements, for example, in education and career.

Goodhart is of the view that the social division between geographically rooted Somewheres and footloose Anywheres creates a cleavage that is driving populism in Europe, the UK and the USA.[12] I agree that populism is on the rise, but I disagree with Goodhart on who owns the future. His view is that the future belongs to the Somewheres, because the Anywheres' perspective has become over-dominant in recent decades and needs a democratic re-balancing.[13] He also points out that the Somewheres are the majority and they are more politically demanding. The title of his book *"The Road to Somewhere"* also indicates that he thinks the road of demographic development favors the Somewheres. I agree that the Somewheres are currently greater in number and more demanding. The Somewheres exercised their significant political clout and brought a number of populist leaders to power in and around 2016. While the Somewheres had legitimate concerns about the economic and social situations in their countries, their demands turned out to be unrealistic and the populist leaders they elected found it difficult to meet their campaign promises. From the advantage of 2021, I look back and wonder: Did the Somewheres vote for leaders who are capable of satisfying their demands? Were their demands achievable? Were the Somewheres willing to invest their cognitive and psychological resources to achieve their demands, or did they rely on political leaders to satisfy their demands? Is the cleavage due to just politics? I believe this cleavage of the Western populations is not just about politics; it is also about globalization, technology and psychology.

In my reading of Goodhart's book, it appears that the Somewheres have been with us for generations, and the Anywheres are emerging from the population of Somewheres. Geographically rooted, and usually not university educated, the Somewheres tend to vote for populist leaders who they expect will restore past national pride and recover past opportunities for their nation. The Anywheres are usually geographically mobile, university educated and, rather than rely on national leaders to improve their situation, they vote with their feet by going to locations where their expectations can be met. Goodhart notes that Somewheres tend to vote for Conservative leaders, while Anywheres tend to vote for Liberal leaders. The populist leaders who came into power in and around 2016 were elected by Somewheres who outnumber Anywheres in voting power.

2016 Elections in UK, USA and European Countries

I am writing in 2021, so I have the advantage of observing the outcomes of 2016 elections in the United Kingdom (UK) and the Unites States of America (USA), Poland, Hungary and Italy. The BREXIT deal on the withdrawal of the UK from the European Union (EU) was agreed in the last week of December 2020. Although the majority of British voters favored leaving the EU, the BREXIT demands seemed to have been based more on angry feelings about the loss of British autonomy, rather than deliberate thought given to what exit from the EU actually means for British citizens. In her 2016 Conservative conference speech, British Prime Minister Teresa May adopted the stance that: "If you believe you are a citizen of the world, you are a citizen of nowhere. You don't understand what citizenship means." [14] That comment seems to place her among Goodhart's Somewheres.

When Teresa May's proposals for BREXIT Deal were rejected, she resigned.[15] Boris Johnson became Prime Minister. He achieved an agreement in December 2020. It took the British leaders 4 years to work out a deal. Although the UK's exit from the EU was scheduled for January 31, 2021, the BREXIT Deal had gaps regarding trade negotiations, two-way traffic across borders and ports, how Britain will acquire the 30% of its food that used to be obtained from the EU, all of which are to be determined at a later date.[16] The UK will not regain control over its own the fishing waters until 2026. Decisions about storage, access and processing of law enforcement databases are also to be determined.[17]

It took four years to work out the BREXIT Deal and the implementation began in February 2021, but the restoration of past British glory is not yet in sight. Where that restoration depends on rolling back globalization or technology, it is unlikely to occur. The economic viability of BREXIT has to be worked out. Will the Europeans continue to regard London as a financial center? Will departure from the EU restore Britain's sovereignty? Will nationalism improve the UK's standing in the financial world? If the UK does better than the EU in stemming the tides of immigrants from conflicts in the Middle Eastern countries, will it restore Britain to pre-EU glory? Will the UK be able to attain an immigration pattern that relinquishes a weak European identity and restores a strong British identity? Answers to those questions are still being debated.

The 2016 USA election brought a populist leader to power on promises to "Make America Great Again" ... a reference to glories of the past. One prominent campaign promise was that immigration would be curtailed by the building of a wall along the length of the southern border and that Mexico would pay for the wall. That has not happened. It was unrealistic to imagine that Mexico, a sovereign country, could be forced to pay for the building of a wall. Another popular campaign promise was that the Affordable Care Act would be repealed and replaced. There was an attempt to repeal, but it faltered because no replacement had been prepared. A third popular campaign promise was that political corruption would be addressed by draining the swamp in Washington. Instead, several people in the new administration were convicted, and some sent to prison for unlawful behavior. There were campaign promises that factories and jobs which had been moved overseas in previous administrations would be brought back to the USA. There was no significant movement of factories returning to the USA. There was no turning back of the technology that had replaced manual jobs.

In addition to the populist leaders in the UK and USA, there were European countries that had populist leaders between 2016 and 2020, for example, Poland, Hungary and Italy. Their populist demands were also based on legitimate concerns of citizens, but were as difficult to satisfy. The web site for the Tony Blair Institute for Global Change offers the observation that populist administrations offer promises that are unfulfillable:[18]

> "This series begins from the understanding that populism often arises from serious and legitimate concerns about the quality of institutions and political representation in countries. ... Populism's appeal is often based on real concerns about the failure of mainstream parties to address issues that citizens are worried about and the failure of institutions to deliver policy outcomes that matter to citizens. Populism can also arise in contexts of profound economic failures, where economic systems do require disruptive transformation to deliver broad-based growth.
>
> It is often lamented that populism threatens to destroy independent and objective institutions that are essential to well-functioning democracies. Yet all too often, by the time populism arises, these institutions—like the media, the judiciary and independent governmental agencies—have long not been working as promised. Populists break onto the scene by pointing to these flaws in the established political system—flaws that mainstream parties may have been sweeping under the carpet for years—and promising far-reaching solutions. Raising political questions that have been too long depoliticised and promising institutional reforms are necessary and important initiatives that political leaders should undertake. The problem with populists is that they raise these issues as a means of riling their base and dividing societies. The solutions they promise, however, are fantasies, characterised by vague ideas and unfulfillable promises."

I believe the populist leaders tapped into legitimate concerns of their fellow citizens. They tapped into the gross dissatisfactions of portions of their populations that felt left behind and tried to address the issues. I think that the Somewheres of the UK and USA wanted to restore their countries to past glory. Isolationism. Protectionism. Strong unions. Nostalgia for privileges, opportunities and benefits that were available in the 20th century. Restoration of past glory is a naïve aspiration, because the circumstances of the 20th century no longer exist. We are now in the 21st century when the forces of globalization and technology prevail. Those are widespread forces. They are the phenomena of evolution, and I do not think they can be restrained by individual nations. In their anger about the present, Somewheres tried to return to what they had experienced as better times in the past. That sentiment is understandable. However, we cannot turn back time, or globalization, or technology. Because individual nations cannot turn back time, or globalization or technology, I would say the populist leaders were at a serious disadvantage in trying to satisfy the demands of the Somewheres. The demands were not realistic. Satisfying the nostalgic yearnings for circumstances of the 20th century are not achievable goals for the 21st century.

Cognitive Ability Spectrum

It is unclear whether the Somewheres were willing to invest their own cognitive and psychological resources to achieve their demands. That is because the focus has not been on Somewheres applying their resources to achieve their demands; the focus has been on getting government agencies and institutions to do the work of satisfying their demands. The title of his book – "*The Road To Somewhere*" – indicates Goodhart's expectation that the future belongs to the Somewheres. I disagree. While I believe the Somewheres deserve all the privileges and benefits that go with citizenship, they have an outlook that differs markedly from that of the Anywheres. Goodhart characterizes that as a difference in placement along the "cognitive ability spectrum".[19] I would add that there is also a difference in placement along a psychological spectrum, which I will describe later in this chapter. Before I address Goodhart's cognitive ability spectrum, I want to point out that cognition occurs at two levels. There is cognition, which is about knowing, and there is metacognition, which is about awareness and control of how we know. This is how the Britannica web site defines cognition:[20]

Cognition is made up of "the states and processes involved in knowing, which ... include perception and judgment. Cognition includes all conscious and unconscious processes by which knowledge is accumulated, such as perceiving, recognizing, conceiving, and reasoning. Put differently, cognition is a state or experience of knowing that can be distinguished from an experience of feeling or willing."

This is how the ScienceDirect web site defines metacognition:[21]

"**Metacognition** is 'cognition' about cognition, 'thinking' about thinking, 'knowing' about knowing, becoming 'aware of one's awareness' and higher-order thinking skills. The term comes from the root word meta, meaning "beyond", or "on top of". Metacognition can take many forms; it includes knowledge about when and how to use particular strategies for learning or problem-solving. There are generally two components of metacognition: (1) knowledge about cognition and (2) regulation of cognition."

Because Goodhart ties cognition to meritocracy, he limits cognition to the formal aspects of knowledge acquisition, that is, the parts of cognition that formal education measures. Those types of knowledge acquisition and measured credentials serve useful purposes, but they are mostly about the content of one's knowledge. They take little account of the awareness and management of one's knowledge, that is, metacognition. Goodhart uses the expression "cognitive ability spectrum" as a reference to mental skills on a continuum that stretches from the most capable to the least capable in a given population.[22]

He ties cognitive ability to meritocracy for the purpose of pointing out that while meritocracy does provide a ladder for social mobility, it has limitations:[23]

"As societies become more mobile and less caste or class based so do differences in a cognitive ability between people become more salient ...

Meritocracy is unassailable in principle but … in practice it can legitimize inequality and reduce empathy for the poor …

Human beings are group creatures and the upwardly mobile, like the immigrant, voluntarily relinquishes the security of the group for the advantage of belonging to a higher social class, or in the case of the immigrant, more successful country."

Goodhart assumes that a person who differentiates themself from their community of origin is necessarily at a disadvantage in a meritocracy. He does not take psychological maturity into account in scaling the ladder of meritocracy. Although he acknowledges that self-realization and autonomy are influential, he does not seem to consider that the effort involved in self-realization fosters the strong sense of agency and the clear definition of one's identity that underpin the scaling of the ladder of meritocracy.

Goodhart states:[24]

"And the growing centrality of educational attainment to the allocation of high-status jobs – combined with a dominant assumption about the virtues of meritocracy and upward social mobility – has made it more likely that the people in the bottom half of the income spectrum and the cognitive ability spectrum will now feel unsuccessful rather than merely unlucky or unambitious."

He further states:[25]

"Social mobility and meritocracy are central to the Anywheres' progressive individualist outlook but Somewheres have some cause to feel more ambivalent about them. They are based on the unspoken assumption of an achieving society, and are about ambition and success as well as about fairness. There is nothing wrong with ambition and success – indeed a successful society needs to give a high priority to both principles and the people who pursue them. But what about everybody else?

> Advocates of social mobility too rarely pause to consider the effect on those who do not climb the ladder – and ... half of the population is always, by definition, in the bottom half of the income and cognitive ability spectrum. In a more individualistic and competitive society we are valued, at least in the public sphere, by what we achieve rather than who we are, creating a constant threat of low esteem for the less successful. This is an inevitable aspect of modern life but it sets up a tension with that more egalitarian promise of a citizen's entitlement to security and a decent life – perhaps even to recognition and esteem."

Goodhart's cognitive ability spectrum is tied to meritocracy of educational attainment. He informs us that one factor which distinguishes Anywheres from Somewheres is the former being more inclined to acquire university education. In my view, a university education was a major contributor on the path to success and satisfaction with life in the 20th century. In the 21st century it is not necessary for success. What is necessary in the 21st century is a marketable skill, regardless of whether that skill derives from formal educational credentials. Many of the successful entrepreneurs of the 21st century either did not go to university, or dropped out. University education has become prohibitively expensive, it consumes multiple years, and it does not guarantee success. The availability of the Internet offers many more opportunities now than were available in the 20th century. Online training. A platform for advertising products and services. Social media for engaging clients. Independence to choose careers that are not derived from educational credentials. Autonomy to differentiate one's branding from the competition. Upward mobility based on attainment of formal educational is chartered territory, while upward mobility based on the use of Internet facilities is unchartered. To scale the ladder of upward mobility via the use of technological opportunities involves being willing to become aware of one's cognition and being able to take control of it. Based on Goodhart's distinction, the Anywheres are better at demonstrating the ability to control their cognition.

If the Somewheres and the Anywheres both took advantage of the same opportunities available in the meritocracies of Western countries, and the Somewheres got less rewards than the Anywheres, that would be justification for an argument that meritocracy has a causal influence in the Somewheres being left behind. However, that is not what is happening. The Somewheres feel left behind by globalization and technology, both of which require heightened cognition. The cognitive abilities that were enough to carve out a successful path in life during the 20th century are different from the cognitive abilities necessary for a successful life in the 21st century.

Meritocracy

In his book "The Meritocracy Trap" law professor Daniel Markovits indicates that he sees meritocracy as a trap that torments both the middle class and the working class.[26] In his opinion, those who meritocracy traps into upward mobility are mercilessly dragged into an unending spiral of achieving merit followed by longer working hours necessary to stay competitive. Markovits regards those who meritocracy traps into downward mobility as being propelled into the arms of populist leaders who are incapable of addressing their concerns about being left behind. Markovits does not use the words "Somewheres" and "Anywheres" but his work is relatable to Goodhart's populations because Goodhart characterizes Somewheres as inclined to be Conservatives and Anywheres as inclined to be Liberals. Markovits also describes the impact of meritocracy in terms of Conservatives and Liberals:[27]

> "Demagogues (populist leaders) inflame middle-class resentment by railing against a corrupt establishment and attacking vulnerable outsiders. They promise, through these attacks, to restore a mythical golden age. President Trump says that ... deporting undocumented workers and families will Make America Great Again. Nigel Farage argues that closing the border to the European Union will restore Britain's independence and

self-respect. And German populists, seeking to recover 'a thousand years of successful German history,' accused Angela Merkel of betraying her country by admitting refugees.

...

(Those who voted for populist leaders) remain under meritocracy's thumb. They are captives who embrace their captor through a sort of ideological Stockholm syndrome.

...

The middle-class and the elite are differently tormented, but by the same oppressor. To (escape), they must dismantle meritocratic hierarchy and build democratic equality – a social and economic order that serves everyone and in which the status of each is valuable precisely because it is shared by all."

Markovits sees two paths to dismantling the meritocracy trap; one is education reform and the other is workforce reform:[28]

- **Education reform**: Make education inclusive by opening the pipelines through which schools and universities offer social mobility. Replace extravagant, competitive training with education that is more open and inclusive. Stop treating universities as charities, since they charge enormous fees. They should be treated as charities only if they function openly and inclusively. Greater openness would increase social mobility.
- **Workforce reform**: Rebalance production away from superordinate workers toward middle-class labor. Promote middle-class labor by producing goods and services that favor mid-skilled workers. Corporate governance and government regulations currently influence the type of jobs that exist and how they are rewarded and how they are taxed. Replace gloomy and glossy jobs with mid-skilled labor that is at the center of economic production. Any industry that is concentrated in a

superordinate working class must be dispersed across a broad middle-class. An increased supply of educated workers would reduce the excessive work hours that elite jobs require for professionals to maintain their positions.

Markovits acknowledges that these ideas are not instructions for curing the flaws of meritocracy, and he is aware that these ideas would be an effort that would be generational. That is understandable, but then he goes on to propose that innovators would have the incentive to "bend the arc of innovation" away from technologies that favor superordinate workers toward those that favor the working-class.[29]

His idea for improving meritocracy is to make education inclusive by opening the pipelines through which universities offer social mobility. In my observation, universities do not actually offer social mobility; universities offer degrees. Social mobility is about what graduates do with their degrees. In my opinion, graduates with greater metacognition, a stronger sense of agency, and more willingness to define their own identity, demonstrate greater innovation in climbing the ladder of meritocracy.

Yes, technology does bias the labor market, but it is unrealistic to image that the "arc of innovation" can be bent away from technology. Following their elections around 2016, populist leaders of the UK, USA and European countries have not bent the arc of innovation away from technology. They have not turned back technology. I believe that the unfolding of technology is an evolutionary phenomenon of life in the 21st century, in the sense that humans observe it and participate in it, but technology is not dictated or controlled by humans. We do not control technology any more than we control evolution. What is more likely is that technology will take over the excessive work being demanded of superordinate workers. That will free up humans to attend to the humane, compassionate, and creative aspects of business models.

A redefinition of meritocracy is potentially useful, but technology is not responsible for our human failure to sustain a meritocracy that values creative skills as much as economic skills. The problem may not be with meritocracy itself, but with what we value as a Western society. If the Western society rewarded creative skills as much as we reward economic skills, meritocracy would be more balanced. So, we need to address the question: Why does our Western society value some skills more than others. Do we assign greater value to those skills that enable our society to thrive?

Markovits does have a point that meritocracy is being rigged by wealthy parents who pay for their children to go to Ivy League universities. But that does not explain the success of people who do not go to university, or people who do not have wealthy parents. Here are examples of two groups of people who undermine Markovits' argument:

- Entrepreneurs who succeeded without university degrees.
- Deferred Action for Childhood Arrivals (DACA) participants who are the offspring of immigrants.

Some of the most productive people in the 21st century had the opportunities to benefit from meritocracy, but chose their autonomous inclinations. Autonomously, against the advice of their parents and mentors, they decided to drop out of university. For example, Bill Gates, Steve Jobs and Mark Zuckerberg all dropped out of university. Each of them felt a psychological connection with a particular technology wave and chose to ride it while it was cresting. They cultivated the innate characteristics of pioneers. The meritocracy did not imbue them with characteristics of innovation.

In the 21st century, university degrees still have value for getting in the door of traditional employers such as banks and accounting firms, but are no longer the major item needed for social or economic success. What I view as necessary for economic success are a marketable skill coupled with the cognitive abilities to carve out a path in life for one's self. Many of the more successful passed over the opportunity that the meritocracy offered to obtain a university degree. The web site "100 Top Entrepreneurs Who Succeeded Without A College Degree" displays one hundred entrepreneurs who established successful careers without relying on university degrees. I selected this list to show the variety of industries where success has been achieved without relying on the credentials of the educational meritocracy:[30]

- **Rachael Ray**, Food Network cooking show star, food industry entrepreneur
- **Sean John Combs**, entertainer, producer, fashion designer, and entrepreneur
- **Steve Madden**, shoe designer
- **Richard Branson**, founder of Virgin Records, Virgin Atlantic Airways, Virgin Mobile
- **Simon Cowell**, TV producer, music judge, American Idol, The X Factor, and Britain's Got Talent
- **Barbara Lynch**, chef, owner of a group of restaurants
- **James Cameron**, Oscar-winning director, screenwriter, and producer
- **John Mackey**, founder of Whole Foods
- **Kenny Johnson**, founder of Dial-A-Waiter restaurant delivery
- **Michael Dell**, founder of Dell Computers
- **Debbi Fields**, founder of Mrs. Fields Chocolate Chippery
- **David Neeleman**, founder of JetBlue airlines
- **Bill Gates**, co-founder of Microsoft
- **Steve Jobs**, co-founder of Apple
- **Hyman Golden**, co-founder of Snapple
- **Mark Zuckerberg**, co-founder of Facebook, and
- **Andrew Perlman**, founder of GreatPoint Energy, an Internet communications company.

These entrepreneurs did not rely on wealthy parents, or seek Ivy League university degrees. They escaped the meritocracy trap by carving out an identity for themselves, relying on their innovations, developing their sense of autonomy, and honing their native cognitive abilities. Goodhart would label them Anywheres because they are futuristic in their out-look, they do not yearn for the past and they do not wait for political leaders to create opportunities. They create their futures by developing innovative products and services which find resonance with the public. They demonstrate the metacognition necessary to carve out innovative roles in life.

Examples of people who establish successful careers by taking advantage of the meritocracy, but without wealthy parents financing their university education, are people in the Deferred Action for Childhood Arrivals (DACA) program in the United States of America.[31] It is a program that allows approximately 800,000 undocumented immigrants, who came to the USA illegally as children, to get work permits and deferral from deportation. DACA recipients do not have wealthy parents paying for their education. They are not eligible for federal student aid and are barred from public school in some states. Their parents were refugees from South American countries, where their safety was threatened by gangs and political corruption. Several got financial help from TheDream.US,[32] a privately funded organization that helps refugees to attend the more affordable universities. Overall, many DACA recipients have carved out successful lives. The significant contributions they are making to society became noticeable in 2017, when the Trump administration announced that it would rescind the DACA program. Business leaders in several states raised objections by pointing out that the loss of DACA recipients would have a negative impact on their industries. The tech giants in Silicon Valley were especially vocal in their objections to the DACA program being rescinded.[33]

DACA employees make a significant contribution to the industries in which they work. In the face of great uncertainty and with little support, they constructed identities for themselves and developed meta-cognitive skills adequate to sustain themselves emotionally and establish careers to support themselves financially. They are successful, but it is not because of wealthy parents. Their success comes from an inner drive to succeed in spite of living in situations of great uncertainty.

They have no permanent home in the USA, they experience repeated threats of deportation, and they are unable to make lasting plans for the future. Despite living in a continuous state of being undocumented and on the brink of deportation, they acquired marketable skills, some by formal education, others by self-taught skills. Some were able to find opportunities in the technology industry by teaching themselves to write computer programs. DACA employees provide examples of people who succeed without wealthy parents to fund their education. The success of DACA recipients demonstrates that success is not just the result of the meritocracy derived from the structure of educational and corporate institutions. Success is also about people having the motivation and the drive to exert effort necessary to achieve their goals. The DACA people escaped the meritocracy trap by their sense of autonomy combined with a carefully constructed identity, and the development of cognitive abilities. Goodhart might call the DACA people Anywheres because they are not geographically rooted, they demonstrate a sense of autonomy, they put effort into self-realization, and they exhibit strong cognitive abilities. Their success is seen in the volume and diversity of industry leaders who protested the Trump administration's decision to rescind the DACA program.

Entrepreneurs and DACA people succeed, not because they rely on meritocracy and wealthy parents, but because they cultivate cognitive abilities. Goodhart's observations about cognitive abilities are backed up by those of historian Yuval Noah Harari. Historically, cognitive abilities have been drivers in the evolution of Homo Sapiens.

Harari on the Cognitive Abilities of Homo Sapiens

Historian Yuval Noah Harari points out that Homo Sapiens survived while all other human species became extinct. He attributes the success of Homo Sapiens to the acquisition of cognitive abilities during the Cognitive Revolution. These cognitive abilities include the acquisition of a new language, the invention of sophisticated tools, and use of the imagination to compose stories that served as collectively shared mental mechanisms for establishing social norms among tribes.[34] He refers to these products of the imagination as fiction and myth.

This is how Harari describes the cognitive abilities that enabled Homo Sapiens to survive:[35]

"(A)bout 70,000 years ago, Homo Sapiens ... drove ... all other human species ... from the face of the earth. The period from about 70,000 to about 30,000 years ago witnessed the invention of boats, oil lamps, bows and arrows and needles (essential for sewing warm clothing). The first objects (of) art date from this era ... as does the first clear evidence for religion, commerce and social stratification.

(R)esearchers believe that these ... accomplishments were the product of a revolution in Sapiens' cognitive abilities.

The appearance of new ways of thinking and communicating, between 70,000 and 30,000 years ago, constitutes the Cognitive Revolution. The most commonly believed theory argues that accidental genetic mutation changed the ... brain of Sapiens, enabling ... an altogether new type of language. ... What was so special about the ... language that it enabled us to conquer the world?

The ... common answer is that our language is amazingly supple. We can connect a limited number of sounds and signs to produce an infinite number of sentences ... We can therefore ingest, store and communicate a prodigious amount of information about the surrounding world.

(T)he most important information that needed to be conveyed was about humans ...

The ...unique feature of our language is ... the ability to transmit information about things that do not exist ... Legends, myths, ... and religions appeared ... with the Cognitive Revolution. ... The ability to speak about fictions is the most unique feature of Sapiens language.

(F)iction ... enabled us not merely to imagine things, but to do so collectively. We can weave common myths such as the biblical creation story, ... and the nationalistic myths of modern states. (M)yths give Sapiens the unprecedented ability to cooperate flexibly in large numbers."

According to Harari, Homo Sapiens used new intelligence to invent tools and to create social bonds of culture that enabled Sapiens to outmanoeuvre other species and to thrive. Sapiens created myths that were shared collectively among socially connected communities. Although the Homo Sapiens species was not as physically strong as the Neanderthals, Sapiens' superior cognitive abilities gave them the advantage in establishing communities and enforcing social norms.

Harari ascribes Homo Sapiens' survival to evolving cognitive abilities that produce myths which are collectively shared. Goodhart places human ability to thrive on what he calls a cognitive ability spectrum, that is, a continuum of human values collectively shared across a population. To escape the meritocracy trap, Markovits recommends dismantling the meritocratic hierarchy of predefined cognitive abilities and building a social order that serves the entire population because it is collectively shared. Harari looks at collectively shared norms through the lens of history. Goodhart applies the lens of demography. Markovits' outlook is through the lens of law. Demographer Goodhart sees cognitive ability as a differentiator of populations, a marker for upward social mobility and the propensity to benefit from meritocracy. Historian Harari sees cognitive ability as the differentiator of species, the driver of evolution, the differentiator that separated Homo Sapiens from the species that became extinct.

Harari, Goodhart and Markovits are all writing about people, but none of them addresses the psychology of the people or the populations involved. That is probably because our Western education system calibrates knowledge into isolated disciplines, such as history, demography, law and psychology. With respect for the knowledge of Harari, Goodhart and Markovits, I take the liberty of suggesting that we look at what I call the psychology spectrum, as an analogue to Goodhart's cognitive ability spectrum.

The Psychology Spectrum

Goodhart does not mention anything like a psychology spectrum. However, he does remark on the Anywheres being more inclined to autonomy and self-realization than the Somewheres.[36]

This is how the online Glossary of Psychology defines autonomy:[37]

> "Autonomy refers to the capacity to make decisions independently, to serve as one's own source of emotional strength, and to otherwise manage one's life tasks without depending on others for assistance; an important developmental task of adolescence."

Self-realization appears to be more illusive to define. Here are some definitions of self-realization.

From the Web site for Semantic Scholar:[38]

> "**Self-realization** is a complex process that needs to be addressed from a number of perspectives, to provide a ... true picture of how individual development takes place. ... Humanistic psychology has provided a series of fundamental theories about human personality and its development. Prominent representatives such as A. Maslow, C. Rogers or R. Assagiolli, along with the psychoanalyst C. G. Jung, have defined the basic concepts

that help us today to better understand the individual evolution-ary path from intuitive thinking structures and primary group integration, to elements of metacognition, creativity and inte-gration into society."

From the Web site for the Merriam-Webster Thesaurus:[39]

"Self-realization: The act of achieving the full development of your abilities and talents."

From the Web site of Ego Science:[40]

"Self-realization is about discovering who we truly are out-side of the ego's patterns and habits. This involves letting go of attachments that keep us stuck in old unhelpful cycles, and realizing the power of the mind to re-train and re-direct itself."

As an analogue to the cognitive ability spectrum, I offer the concept of a psychology spectrum, that is, a continuum of individual development across Western populations. Based on the definitions above, I think be-haviors that demonstrate autonomy and self-realization point to greater psychological maturity than those behaviors that do not. Autonomy and self-realization are characteristics that foster development of a sense of agency and a definition of identity, which are necessary to find one's own way in life. Autonomy and self-realization enable a person to set a direction in life and carve out a path of what is possible, given one's po-tential for fulfillment. This is not just for the purpose of earning income, but also for overall satisfaction with life.

The Anywheres demonstrate a sense of agency by directing their own futures. They do not rely on political leaders to create a future for them. The Anywheres define their identities in terms of their achievements, rather than in terms of their familial context. The Somewheres appear to dwell in the angst of being left behind. Instead of taking charge of their lives, they put that responsibility on political leaders. Based on Goodhart's depiction of the Somewheres and the Anywheres, I think the Anywheres are better able to make their way in life. They are not inclined to complain of being left behind. They are comfortable with change. Autonomy and self-realization are notable characteristics of the Anywheres. I think behaviors that demonstrate autonomy and self-realization point to psychological maturation. The behavior of the Anywheres is consistent with the psychological principle of individuation. To support my point that Anywheres are more psychologically mature, I refer to the psychological principle of individuation.

Psychologist Murray Stein wrote "*The Principle of Individuation*" in which he characterizes individuation as a life-long opus based on an innate psychological imperative that aims at increasing consciousness.[41] This life-long effort proceeds in two movements: differentiation and integration.

1. **Differentiation:** The first movement is about breaking down unconsciousness; this involves the differentiation and separation of identities that have their contents outside of the individual, for example, identities that include content acquired from being a member of a family, a community, a religious organization, or a political party. This movement of differentiation enables the individual to cultivate a more lucid consciousness, that is, increases the ability to control events consciously.[42]
2. **Integration:** The second movement is about paying sustained attention to the images that emerge from the unconscious realm of the psyche and integrating the contents into consciousness. Examples are newly emerging images related to dreams, active imagination and synchronistic events. This movement of integration enables the individual to function in a more expansive way during everyday living.[43]

The two movements, differentiation and integration, occur in tandem with each other. Both are necessary for individuation. Stein states:

"(T)he principle of individuation defines something essential about the human being. It is an absolutely fundamental drive in the human subject to distinguish itself from one's surroundings. This is individuation, at least in part, and the energy for its creation is a given of human consciousness. In becoming a person, one must necessarily create distinctions and separateness. The drive to create specificity in human consciousness … is grounded in nature. It is in accord with human nature, therefore, to seek individuation. The movement toward individuation is not optional, not conditional, not subject to the vagaries of cultural differences. It is a given, although of course many people ignore it, repress it, and distort themselves in convoluted attempts to avoid acknowledging its presence out of fear of appearing non-conformist, or being seen as 'different'." [44]

Just as the human body matures by changes over a lifetime, so too, the human psyche matures psychologically over a lifetime. The difference is that the biological maturation is visible, while the psychological maturation has to be inferred from behavior. Goodhart's observation that the Anywheres are more inclined to autonomy and self-actualization seems to me to be an indication that the Anywheres have developed a greater sense of their own agency. The Anywheres seem to have an intuitive grasp of the psychological principle of individuation, and it imbues them with a strong sense of agency, as well as a sense of identity that is not tethered to circumstances of their birth.

In 2016, the Somewheres voted for populist leaders in the UK and the USA. Leave the EU and restore Britain's glory. Make America Great Again. Four years later, the Somewheres are still angry because many campaign promises were not kept. The point made on the web site for the Tony Blair Institute for Global Change is that populist administrations offer promises that cannot be fulfilled. The 2016 populist wins in UK, USA and European countries gave voice those who are angry about being marginalized, and who resent the educated elite. However, they have not gained much because they surrendered their voting power to leaders who do not match their needs psychologically. They want leaders who will create a psychological container for them, bolster their sense of self-importance by turning back globalization and technology, which are two major contributors to their sense of being left behind. Four years later, populist leaders show no signs of being effective psychological containers, little inclination to bolster their voters' self-importance, and no appetite for turning back either globalization or technology. The Somewheres are still angry. They are still being left behind.

The Anywheres are not being left behind. They do not rely on political leaders to create a psychological space for them to function; they create their own space. They are geographically mobile. They are unencumbered by the geographic circumstances of their birth. Untethered to the community of their birth. They have crafted identities not defined by geography. They acquired skills that are mobile, and can be practised anywhere. Most important, they have cultivated a sense of their own identity that eliminates dependence on political leaders.

The Future Belongs to the Anywheres

Goodhart believes the future belongs to the Somewheres because they outnumber the Anywheres and therefore have the voting power to choose political leaders. The wording in the title of his book "The Road To Somewhere" reflects his view that the Somewheres have the upper hand and that the road to the future leads to Somewhere. It is true that the Somewheres outnumber the Anywheres at present, but the number of Anywheres is growing while the number of Somewheres decreases. It is also true that the Somewheres have the volume of voters to put in power the leaders of their choice. Voters in the UK and USA put their voting power on display in 2016. It appears that a vociferous show of voting power does not translate to ability to choose effective leaders. Four years after the BREXIT vote, a deal has been agreed, but it has many gaps. Four years after the 2016 USA election, the anger of the populists amplified into an assault on the USA Capitol in the hope of overturning an election that had been certified in all 50 states by governors of both major parties.

I believe the future belongs to those who are more psychologically mature. The Anywheres place a high value on personal identity, autonomy, mobility and novelty, but a lower value on group identity, tradition, and national norms.[45] I see the Anywheres as owners of the future. The Anywheres have a well-developed sense of their own agency, so they do not need to depend on politicians. The Anywheres' mobility and achievements equip them for future prospects, so they are not as inclined to look backward and yearn for the past as the Somewheres do. The Anywheres do not depend on political leaders to bring back privileges from the past. They do not wait for a return to the Britain that existed before joining the EU. They do not wait for politicians to revive dying industries and Make America Great Again. The Anywheres carve out their own future with and without the formal meritocracies of the Western world. The Anywheres are comfortable with the progress of globalization and technology, both of which I consider to be the drivers of the future. Another driver is psychological maturity that comes with the recognition that one's growth necessitates differentiation from identities with family and community.

Comparing Goodhart's Trend In Demography With Giegerich's Trend In Psychology

To further explain the bifurcation of identity that splits the Anywheres from the Somewheres, I compare Goodhart's trend in demography with Giegerich's trend in psychology. I describe Giegerich's trend in psychology in Chapter 2: "Wolfgang Giegerich's Trend in Psychology".

Here is my summary of Giegerich's trend.

There is a paradigm shift ...

From a focus on the semantical level of psychology,
where individuals engage in the individuation process,
a goal-seeking effort to differentiate their minds from the
unconsciousness of their communities,

To a focus on the syntactical level of psychology,
where human culture engages in the interiorization process,
an intellectual discipline of interpreting phenomena
that emerge in the world.

The Somewheres and the Anywheres differ significantly in their sense of identity, their outlook on life, and their aspirations. The sense of identity revealed by the behaviors of the Somewheres is so different from the Anywheres, that I believe the emergence of the Anywheres involves a restructure of consciousness. The transition from a culture of Somewheres' ascribed identity to a culture of Anywheres' achieved identity involves such a dramatic change of outlook on life that it implies a corresponding restructure of consciousness. This is consistent with Giegerich's trend which is about movement from the semantical level of psychology, to the syntactical level of psychology. As Goodhart points out, the two groups are widely separated on the cognitive ability spectrum. Because the Anywheres are more receptive to rapid changes in globalization, ethnic mixing of society, and advancements in technology, I see them as functioning at a higher level than the Somewheres. Accommodating rapid changes in all those areas of life necessitate a restructuring of consciousness. To relinquish an identity defined in terms of heritage, and embrace an identity defined in terms of achievements necessarily involves a restructure of consciousness. This chapter shows that Goodhart's trend in demography and Giegerich's trend in psychology are discipline-specific aspects of the same paradigm shift. Goodhart views the paradigm shift in terms of demography, while Giegerich sees the paradigm shift from the point of view of psychology.

What Goodhart sees as a cleavage in Western populations, and Giegerich views as a shift in levels of psychology, are in my opinion two aspects of the same overarching paradigm shift. In essence, the traditional identity of Western culture is splitting into two identities. One defined by heritage, the other defined by accomplishments. In later chapters, I describe additional bifurcations in other disciplines, to indicate an overarching paradigm shift which places Homo Sapiens on the cusp of generating a new version of humanity.

Summary

Identity appears to be central to the cleavage of Western populations into Anywheres and Somewheres. How people define their identity is one of the factors that distinguish David Goodhart's Somewheres from Anywheres. According to Goodhart, Somewheres have an ascribed identity, while Anywheres have an achieved identity. I propose that the demographic cleavage into groups of Somewhere and Anywheres is a bifurcation of identity of the populations of Western countries. Goodhart's trend in demography begins with the stable Somewheres, who have a history of taking pride in belonging to their local community, and the trend progresses to Anywheres, who are newcomers more interested in what globalization has to offer. Goodhart sees cognitive abilities as a significant differentiator between the Somewheres and the Anywheres. He places Somewheres low on his cognitive ability spectrum, and places the Anywheres high on the spectrum. I would add that there is also a difference in placement along a psychological spectrum, a difference in being able to craft one's identity.

Because Goodhart ties cognition to meritocracy, he limits cognition to the formal aspects of knowledge acquisition, that is, the parts of cognition that formal education measures. Those types of knowledge acquisition and measured credentials serve useful purposes, but they are mostly about the content of one's knowledge. There take little account of the awareness and management of one's knowledge, that is, metacognition. People who have strong metacognition are not encumbered by the lack of university degrees, and they do not need the containment of an employer to become successful. Earlier in this chapter, I quoted from a list of an increasing number of people who achieve successful careers, without university degrees. I think these people demonstrate greater metacognition than many of their university credentialed counterparts. James Cameron. Rachel Ray. Bill Gates. Simon Cowell. Richard Branson. These are people who bypassed the opportunity to take the traditional path of obtaining a university because they sought opportunities better suited to their sense of their identities.

To make sense of one's experience, a person constructs an autobio-graphical narrative from childhood through adolescence into adulthood. The autobiographical narrative that a person constructs is the basis of their sense of identity. That construction occurs in conjunction with accumulation of cognition throughout life, whether that be acquired in educational institutions, at the workplace, or elsewhere. Cognition is enhanced by metacognition, an awareness of one's own thought pro-cesses. Metacognition is about being able to assess and monitor one's social skills and intellectual skills. It is also about recognizing the limits of one's ability. Metacognition involves the management of the content of one's cognition to regulate one's performance and mental processes. Well-developed metacognition makes a person more inclined to take the approach that circumstances are within their control.

While Goodhart's cognitive ability spectrum does not explicitly mention metacognition, it is implied in the differences between his characteriza-tion of the Somewheres having an ascribed identity and the Anywheres having an achieved identity. The Somewheres' ascribed identity has the quality of being inherited, of being passed on from family and communi-ty, of being similar to the previous generation. The Anywheres' achieved identity has the quality of being earned, of being the accomplishment of individuals, of being different from previous generations. I propose that the emerging of the group of Anywheres from the group of Somewheres entails a bifurcation of identity in Western populations.

This chapter explains my observation of a bifurcation of identity into ascribed identity and achieved identity. That bifurcation, together with the bifurcations I describe in other chapters, point to an overarching paradigm shift that indicate Homo Sapiens on the verge of generating a new version of humanity.

The next chapter is about Gartner's trend in business operations. Since the Anywheres are comfortable with changes brought about by tech-nology, I expect that they are the people populating the "Anywhere Operations" which is one of the trends identified by Gartner. In the next chapter, I describe Gartner's trend in business models from (Somewhere) Operations to Anywhere Operations.

NOTES:

1. *"The Road to Somewhere: The Populist Revolt and the Future of Politics"* by demographer David Goodhart.
2. *"Populists in Power Around the World"* from the web site of the Tony Blair Institute for Global Change.
3. *"The Meritocracy Trap"* by law professor Daniel Markovits.
4. *"The Individuation Principle"* by psychologist Murray Stein.
5. *"Sapiens: A Brief History of Humankind"* by historian Yuval Noah Harari.
6. See the credential of David Goodhart on the Amazon web site: https://www.amazon.com/Road-Somewhere-Populist-Revolt-Politics/dp/1787382680/ref=sr_1_2?Adv-Srch-Books-Submit.x=45&Adv-Srch-Books-Submit.y=11&dchild=1&qid=1605137523&refinements=p_27%3Adavid+goodhart%2Cp_20%3AEnglish&s=books&sr=1-2&unfiltered=1.
7. See Goodhart's sources in *"The Road To Somewhere"*, p viii; p xv.
8. See *"The Road To Somewhere"*, p vii.
9. See *"The Road To Somewhere"*, pp 3 - 4.
10. See *"The Road To Somewhere"*, pp 3 – 13.
11. See *"The Road To Somewhere"*, pp 3 – 13.
12. See *"The Road To Somewhere"*, p 4.
13. See *"The Road To Somewhere"*, p viii.
14. See the BBC web site: 'Mrs May, we are all citizens of the world,' says philosopher - BBC News.
15. See the BBC web site: Theresa May resigns over Brexit: What happened? - BBC News.
16. See the BBC web site: (Brexit: Are the borders ready? - BBC News).
17. See the BBC web site: (Brexit: What are the key points of the deal? - BBC News).

18. See the article: "Populists in Power Around the World" dated November 2018, on web site: <u>Populists in Power Around the World | Institute for Global Change.</u>
19. See *"The Road To Somewhere"*, p 180.
20. See the Britannica web site defines cognition: <u>cognition | Definition, Approaches, & Facts | Britannica.</u>
21. See the ScienceDirect web site defines metacognition: <u>Metacognition - an overview | ScienceDirect Topics.</u>
22. See *"The Road To Somewhere"*, p 180.
23. See *"The Road To Somewhere"*, pp 180 – 181.
24. See *"The Road To Somewhere"*, p 152.
25. See *"The Road To Somewhere"*, p 180.
26. See *"The Meritocracy Trap"* by Daniel Markovits, pp 271 – 275.
27. See *"The Meritocracy Trap"*, pp 271 – 275.
28. See *"The Meritocracy Trap"*, pp 275 – 279.
29. See *"The Meritocracy Trap"*, p 240.
30. See the web site: <u>100 Top Entrepreneurs Who Succeeded Without A College Degree (elitedaily.com).</u>
31. See DACA web site: <u>Consideration of Deferred Action for Childhood Arrivals (DACA) | USCIS.</u>
32. See TheDream.US web site: <u>TheDream.US Opportunity Scholarship Program Details - Apply Now | Unigo.</u>
33. See web site: <u>Overnight Tech: Silicon Valley blasts Trump DACA decision | Zuckerberg calls it 'sad day' for country | Apple's Cook 'dismayed' | Lenovo settles FTC privacy charges | Google faces blowback over think tank firing | TheHill.</u>
34. See *"Homo Sapiens"* by Yuval Noah Harari, p 20.
35. See *"Homo Sapiens"*, pp 20 – 25.
36. See *"The Road To Somewhere"*, p 5; p 24.
37. See web site for the Glossary of Psychology: <u>Autonomy (psychology-lexicon.com).</u>
38. See the web site for From the Web site for Semantic Scholar: <u>[PDF] The Process of Self-Realization—From the Humanist Psychology Perspective | Semantic Scholar.</u>

39. See web site for the Merriam-Webster Thesaurus: <u>Self-realization Synonyms | Merriam-Webster Thesaurus (merriam-webster.com).</u>
40. See the Web site of Ego Science: <u>Self Realization vs Self-Actualization - Part 1: Maslow's Hierarchy - Ego Science.</u>
41. See "*The Principle of Individuation*" by Murray Stein, p 5.
42. See "*The Principle of Individuation*", pp 5 – 7.
43. See "*The Principle of Individuation*", pp 5 – 6.
44. See "*The Principle of Individuation*", p 8.
45. See "*The Road To Somewhere*", p 5.

FOUR

Gartner's Trend In Business Operations

There is a paradigm shift ...

From a business model of (Somewhere) Operations
where individual store managers enable staff
to work in specific stores,
and serve customers in specific geographic locations,
during specific hours,

To a business model of Anywhere Operations
where computer networks enable staff
to work from anywhere,
and serve customers everywhere,
and activate operations anytime.

Gartner Incorporated, is a research and consulting firm that offers advice about business operations to corporations and government agencies. In this chapter, I describe Gartner's paradigm shift in business operations. The trend is from a business model of (Somewhere) Operations to Anywhere Operations. I insert the word "Somewhere" as a descriptor because Gartner's trend in business operations has much in common with David Goodhart's Somewheres and Anywheres. For a detailed explanation of Goodhart's cleavage of Western populations into Somewheres and Anywheres, see Chapter 3: "David Goodhart's Trend in Demography". The similarity is so noticeable, it is as if Goodhart's Somewheres populate Gartner's (Somewhere) Operations, and Goodhart's Anywheres populate Gartner's Anywhere Operations. What is evident in Gartner's trend is that there is a bifurcation in business operations.

This is my summary of Gartner's paradigm shift in business operations:[1]

There is a paradigm shift …

From a business model of (Somewhere) Operations
where individual store managers enable staff
to work in specific stores,
and serve customers in specific geographic locations,
during specific hours,

To a business model of Anywhere Operations
where computer networks enable staff
to work from anywhere,
and serve customers everywhere,
and activate operations anytime.

The main sources of information for this chapter are:

- Gartner's web site: www.Gartner.com[2]
- Forbes' web site: www.Forbes.com[3]
- Stefanini GROUP's web site: www.stefanini.com[4]
- IBM's web site: www.IBM.com[5]
- "*The Road to Somewhere: The Populist Revolt and the Future of Politics*" by demographer David Goodhart[6]
- Amazon's web site: www.amazon.com[7]
- Barnes & Noble's web site: www.bn.com[8]
- Crown Publishing Group's web site: www.crownpublishing.com[9]

Gartner did not use the expression "Somewhere Operations" ... I coined that expression to indicate where the shift originates, in contrast to where Gartner sees the shift trending, that is, toward "Anywhere Operations". Here is a description of Anywhere Operations quoted from the article "Gartner Top Strategic Technological Trends for 2021" by Kasey Panetta:[10]

"An anywhere operations model will be vital for businesses to emerge successfully from COVID-19. At its core, this operating model allows for business to be accessed, delivered and enabled anywhere — where customers, employers and business partners operate in physically remote environments.

The model for anywhere operations is 'digital first, remote first'; for example, banks that are mobile-only, but handle everything from transferring funds to opening accounts with no physical interaction. Digital should be the default at all times. That's not to say physical space doesn't have its place, but it should be digitally enhanced, for example, contactless check-out at a physical store, regardless of whether its physical or digital capabilities should be seamlessly delivered."

Gartner designates the early part of the trend simply as "Operations" and the later part of the trend as "Anywhere Operations". I use the expression "Somewhere Operations" partly because Somewhere Operations is a convenient counterpoint to Anywhere Operations. Another reason I use the expression is that Gartner's trend has much in common with David Goodhart's cleavage of Western populations into groups of people he calls Somewheres and Anywheres.

There is a correspondence between Goodhart's Somewheres and Gartner's traditional Operations which I choose to label Somewhere Operations. The Somewheres define their identity partly in terms of their attachment to the geographic stability of the community where they grew up, while Somewhere Operations enable employees to work from somewhere, that is, specific geographic locations. There is also a correspondence between Goodhart's depiction of Anywheres and Gartner's Anywhere Operations. The Anywheres value their geographic mobility, while Anywhere Operations enable employees to work from geographically dispersed locations.

Gartner's Anywhere Operations business model enables customers to access business form anywhere and allows the conduct of business at any time, despite customers, employees and suppliers being in geographically separate locations. Gartner's Anywhere Operation business model is a prediction designed to help businesses function successfully in an evolving business climate. Although Gartner outlined the trend before Coronavirus, the Anywhere Operations business model is conducive to doing business in an environment of restrictions imposed to protect public health during the Coronavirus pandemic.

The trend toward Anywhere Operations does not mean that physical brick-and-mortar stores will be eliminated. It means that, increasingly, in-person contact is being replaced by digital capabilities for a seamless delivery of business services. Some of the digital capability is already in place, in the form of contactless check-out in stores where goods are sold. Another example is the delivery of banking services through pre-recorded telephone scripts that guide customers through processes like the cancellation of checks. Customers with Internet access can engage in business at any time, regardless of time zone.

The trend from Somewhere Operations to Anywhere Operations is not an immediate transition. In between those two types of operations, Gartner sees an intermediate Hybrid Operations. I chose bookstores to illustrate these types of business operations:

- Somewhere Operations in Crown Books Corporation, a brick-and-mortar bookstore
- Anywhere Operations in Amazon, an online bookstore, and
- Hybrid Operations in Barnes & Noble, a combination of brick-and-mortar plus online bookstore.

The core areas of business that Gartner uses to distinguish the Anywhere Operations business model are:[11]

- Collaboration and Productivity
- Secure Remote Access
- Cloud and Edge Infrastructure
- Quantification of the Digital Experience, and
- Automation to Support Remote Operations.

These core areas of business of the Anywhere Operations business model are not intended just for allowing employees to work from home, or to interact with customers virtually; Anywhere Operations also creates added value in these five core areas of business.[12] According to Forbes magazine, companies using the Anywhere Operations business models should consider these core areas of business.[13]

Collaboration and Productivity

"Collaboration and Productivity" refers to a core area of business that involves creating an "augmented office" which replicates the physical office, to enable improved collaboration and productivity among employees.[14] An augmented office is a replication of a physical office for the purpose of having employees see each other's availability virtually, assign tasks, organize meetings, and use mobile apps to simulate activity ordinarily conducted in the physical office. As a digital twin of a physical office, an augmented office includes video conferencing tools, that allow employees to see each other's availability virtually. In addition, office managers can tangibly see space occupancy and usage patterns, have full environmental and security control, and rethink efficient use of resources. For client visits, there are digital solutions. By implementing Extended Reality (XR) tools like remote assistance, an organization can share digital models of devices, record real-time analytics, and leave comments in Augmented Reality (AR) while speaking with clients.[15]

To create an augmented office, an organization can combine a digital twin solution, which is an accurate 3-dimensional replica of the work office, with data collected from the Internet of Things (IoT) sensors, and integrate that with the platforms for video conferencing calls and task management tools.[16] Employees can see each other's availability, while office managers can assess space occupancy and usage patterns with full environmental and security control, as they consider efficient consumption.[17]

For employee collaboration with clients, video conferencing can be supplemented with creation of tangible objects, such as hardware sensors. Before the Coronavirus pandemic, one visit to a client's office could solve a number of problems. During the pandemic, effective collaboration requires the implementation of Extended Reality (XR) tools for remote assistance. That allows the sharing of digital replicas of devices, for observing analytics in real time and contributing comments in Augmented Reality (AR) when communicating with the clients. That arrangement enables the building of productive and efficient collaboration around tangible objects and guides clients in times of isolation.[18]

In the Anywhere Operations business model, a workplace consists of a set of platforms, tools and environments that enable employees to deliver work at a distance. Digital workplace platforms tie all these together.[19]

CMSWire describes a Content Collaboration Platform (CCP) as a digital workplace made up of a collection of products, services, tools and environments, which taken together, enables teams to deliver work in a usable, coherent and productive way. A digital workplace integrates all these together for teams whose members may be inside or outside of an organization.[20] CCPs empower people, enabling new productivity, collaboration and efficiency and can drive change in people's work styles and processes, help meet business priorities and grant security and compliance.

CCPs are increasingly important for remote workers because they offer the following features:[21]

- Mobile access to content repositories
- File synchronization across devices and cloud repositories
- File sharing with people and applications, inside or outside an organization
- Team collaboration, with dedicated folders, and
- A content repository, which can be cloud-based or on-premises.

Secure Remote Access

"Secure Remote Access" is a core area of business that combines cloud solutions with firewalls to provide secure access to virtual environments for IT teams and their clients. It also offers additional security for multiple users operating from different time zones.[22] With the trend toward Anywhere Operations, the security needs of organizations are changing to address employees and assets that exist beyond traditional office barriers, and this warrants a security mesh to be built around a person or a thing to ensure timely responses and a standardized security approach.[23]

Cloud solutions powered by firewalls provide this secure remote access to the virtual environment for development teams and clients. Anywhere Operations make it necessary to ensure secure remote access to the multiple users who are in geographically dispersed locations.[24] The nature of security is changing; it is becoming more modular. Employees and assets exist beyond the traditional office perimeter. A new paradigm of security mesh is necessary for persons and assets that need access to the organization.[25]

A cybersecurity mesh is the technology used in the Anywhere Operations business model to ensure safety and security while accessing information in cloud-based applications and distributed data on devices in a geographically distributed environment.[26] With remote working becoming common, organizations function in geographically distributed environments of employees, vendors, partners, and clients. As Anywhere Operations are exposed to cyber threat, flexible and scalable cybersecurity control is becoming a necessity.[27] Cybersecurity mesh addresses the need for security against cyber threats as more assets are currently existing outside the traditional security perimeter of a work environment. A robust cybersecurity mesh enables the security perimeter to be defined around an asset or a person's identity.[28]

A cybersecurity mesh aims to ensure the safety of sensitive data where there are people working remotely from the office environment.[29] A cybersecurity mesh is an IT security infrastructure that does not focus on building a single 'perimeter' around all devices of an IT network. Instead, it establishes smaller, individual perimeters around each device or access point. This creates a modular and more responsive security architecture covering physically disparate access points of the IT network.[30] Where an organization's resources or critical infrastructure are located outside the traditional perimeter, it means that the organization's business data and assets are also outside the organization's physical boundaries, so security infrastructure needs to secure employees working from home.[31]

Cybersecurity mesh is a building block of the security infrastructure in an organization's digital network. In a collaborative environment, business data are highly mobile and need to remain accessible to many different collaborations and secure from unauthorized access. Cybersecurity mesh is adaptable to emerging threats and changing access needs; it can detect threats in real-time and take immediate action to protect an organization's data, devices and operations in ways that password-protection cannot. A cybersecurity mesh helps secure the network by ensuring all the data, systems, equipment and operations are accessed securely regardless of where they might be located.[32] What replaces password-protection is cybersecurity mesh that is integrated into the digital network or platform development.[33]

As Somewhere Operations give way to Anywhere Operations, the traditional perimeter around the boundary of a physical organization gives way to the contemporary perimeter around the boundary of the identity of a person or an asset in a digital network. Multifactor authentication (MFA) is the means for securing the identity of a person or asset in delivering secure remote access.[34] When a person signs on to online accounts, there is an authentication process to determine if the person is who they say they are.[35]

Traditionally authentication was done with just a username and a password. To improve security, online services such as banks, social media, and stores have added a way for accounts to be more secure by "Multifactor Authentication".[36] When a person signs into an online account, they need more than one authentication factor to verify their identity.

Three common types of factors are:[37]

- "Something you know - Like a password, or a memorized PIN.
- Something you have - Like a smartphone, or a secure USB key.
- Something you are - Like a fingerprint, or facial recognition."

Building authentication from multiple factors related to a person's identity, or an asset, provides a modular approach to security and produces timely responses during authentication.

Cloud and Edge Infrastructure

As the business operation trend moves from Somewhere Operations to Anywhere Operations, it becomes necessary to better secure informational assets for the purpose of disaster recovery, and to process data close to the point where the data are generated. "Cloud and Edge Infrastructure" is a core area of business that Gartner defines as cloud solutions which facilitate data processing at or near the source of data generation, in a way that makes cloud storage repository cost-effective.[38] As organizations seek to enable business continuity and ensure disaster recovery, migration to the cloud becomes increasingly common. Tools necessary for application development and project management processes are being moved to the cloud to ensure secure ongoing access to them when needed for operational continuity.[39]

In the interest of operational continuity, sales and marketing teams migrate Customer Relationship Management (CRM) systems to the cloud from their mobile devices. Clients also migrate assets to the cloud to ensure effective business continuity and faster disaster recovery. Cloud and edge infrastructure enable flexibility, transparency and security of an organization's IT assets by establishing programmable cloud solutions that are cost-effective in terms of usage of assets stored in repositories in the cloud.[40]

Edge computing is about decentralizing computing power, by placing it closer to the points where data are generated. The development of edge computing is being driven by rapid deployment of Internet of Things (IoT) projects for a variety of business, consumer and government activities, for example, the creation of Smart Cities.[41] Edge computing offers real-time insights into performance and facilitates actions that are taken in local geographic settings.[42] There are occasions when digital business initiatives create data that are more efficiently processed if the computing power is close to the location where the data are generated. Edge computing solutions enables localized computing power.[43] As an example, IT Infrastructure and Operations (I&O) leaders who manage edge computing solutions take into consideration the

Internet of Things (IoT), which is a source of data generation from sensors and embedded devices. Edge computing serves as the decentralized extension of the organizational networks, cellular networks, data center networks and cloud storage repositories.[44] A decentralized approach to processing data also addresses digital business infrastructure requirements when the volume and velocity of data increase, because the inefficiency of streaming multiple sources of data to a cloud repository also increases.[45] Increase in the volume of IoT devices and the use of cloud computing are increasing the need for edge computing, where cloud providers develop solutions to distribute their cloud capabilities closer to the edge.[46]

Edge computing solutions can take many forms. They can be mobile in a vehicle or smartphone, or they can be static such as when located in a building management solution or a manufacturing plant. Or, they can be a mixture of the two, such as in hospitals or other medical settings.[47] A wearable health monitor is an example of a basic edge solution, because it can locally analyze data like heart rate or sleep patterns and provide recommendations without a frequent need to connect to the cloud.[48] In a vehicle, for example, an edge solution may aggregate local data from traffic signals, GPS devices, other vehicles, proximity sensors and process this information locally to improve safety or navigation.[49]

In support of Anywhere Operations, I&O leaders take the following actions to protect their organizations against cyberattacks:[50]

- Coordinate when and how network devices are connected, to improve chances they will stand stronger against cyberattacks.
- Rally the entire organization to agree on common governance structure for device connectivity to maintain control of the security of the network.
- Create a device certification process for all devices that must be passed before any device is connected to the enterprise network.

In Anywhere Operations, it is beneficial for IT leaders to formulate a strategy for cloud and edge infrastructure, to obtain the business value of new industry developments in the areas of cloud-native platforms.[51]

Quantification of the Digital Experience

"Quantification of the Digital Experience" is a core business area that Gartner uses to highlight the fact that quantification in the Anywhere Operations business model is the baseline for modern organizations.[52] A prerequisite for an Anywhere Operations is a mindset that is oriented to a digital-first, location-independent outlook providing a seamless and scalable digital experience. This mindset involves changes from traditional approaches to technology infrastructure, management practices, security and governance policies, and employee and customer engagement models.[53] This digital-first mindset aims at ensuring the remote workforce has the necessary support. Gartner places a premium on the technology driving quantification of the digital experience as well as automation to achieve this goal.[54] The expansion of IoT (Internet of Things) and AI support several aspects of Anywhere Operations.

Workplace analytics, remote support and digital experience monitoring minimize risk in digital investment. Quantification of the digital experience is essential in Anywhere Operations.[55] Digital transformation is not a one-size-fits-all situation, but according to Gartner, the checklist remains the same for every organization: Digital-first, remote-first; digitally enhanced physical spaces; and distributed business capabilities.[56] In the world of Anywhere Operations, quantification of the digital experience is being accomplished by AIOps. Artificial Intelligence for IT Operations (AIOps) is software that integrates AI & operations for quantification of the digital experience. AIOps produces analytics from operations data also to provide prescriptive and predictive answers in real-time. These insights produce real-time business performance Key Performance Indicators (KPIs), allow teams to resolve incidents faster, and help avoid incidents altogether.[57] AIOps consists of three main steps: Observe – Engage – Act. AIOps continues to process data to detect new anomalies, and these steps are taken in a continuous cycle.

Here are more detailed descriptions of these steps:[58]

- **Performance Analysis (Observe):** This is a task about quantification of observation of the results of processing real-time data from sources such as traditional IT monitoring and log events. In this step, AI algorithms analyze data for the purpose of automatically detecting anomalies in the data. This algorithmic filtering prevents alert fatigue and reduces the workload of IT operation teams as they don't have to do the same work again for similar situations.
- **Experience Management (Engage):** This step is about quantification of the engagement of IT teams to communicate anomalies. AIOps notifies the related IT teams about the anomalies. These teams will be aware of performance issues beforehand and understand the bottlenecks of their applications. Since similar problems are classified together, AIOps tools reduce alert fatigue.
- **Delivery Automation (Act):** This step is about quantification of action on the part of IT teams. AIOps also increases automation level by routing workflows with or without human intervention. It becomes more accurate as it continuously learns from IT team's actions. It can potentially resolve issues before they reach end-users or even before businesses become aware of them.

By integrating Machine Learning algorithms into Anywhere Operations, organizations can use AIOps to perform analytics to address challenges that involve analyzing increasing amounts of data for quantification of their digital experience.

Automation to Support Remote Operations

"Automation to Support Remote Operations" is a core business area that Gartner defines as using Artificial Intelligence (AI) to evaluate, extract, and analyze data to enhance decision-making.[59] Since AI models include algorithms that are capable of learning from data, the algorithms become more accurate as they are fed increasing volumes of data.[60] The automation of business processes enables faster decision-making.

Somewhere Operations – Crown Books Corporation

I use the expression "Somewhere Operations" to refer to Gartner's traditional business operating model which is designed to function according to a centralized infrastructure. A traditional, centralized infrastructure does not have the digital support that Gartner defines in terms of core business areas: Collaboration and Productivity, Secure Remote Access, Cloud and Edge Infrastructure, Quantification of the Digital Experience, and Automation to Support Remote Operations. Somewhere Operations is about the traditional style of managing brick-and-mortar stores, recruiting employees locally and providing services to customers around the location of the store. That style of conducting business focuses on geographic proximity of store, employees and customers. In the Somewhere Operations business model, many of the responsibilities are performed manually. The store manager is responsible for managing the store in compliance with the centralized infrastructure of the organization, for hiring employees who live in commuting distance from the store, and for attracting customers who live in the area. Employees are responsible for being physically in the store during designated hours of work, and for offering in-person service to customers. Customers are able to conduct business only during the opening hours advertised by the store. Suppliers are responsible for delivering supplies at the location of the store, during the store's opening hours. Store managers collaborate with employees, customers and suppliers. Their collaboration yields productive results and they pursue automation of their business processes, but for them, the digital experience is not a priority. The digital experience is not their default.

Crown Books Corporation is the example I choose to explain an organization with a traditional, centralized infrastructure in a Somewhere Operations business model.

Crown Books Corporation

Crown Books Corporation was once a successful brick-and-mortar bookstore that operated in the late 20th century. News articles and company history revealed that Crown Books applied few of the five core areas of business that Garner identifies as supportive of the digital experience in Anywhere Operations. The sources for information about Crown Books Corporation are web sites: Company Histories, and The Baltimore Sun. Crown Books was an organization that functioned according to the traditional, centralized structure that is now known as the Somewhere Operations business model.

Crown Books was founded in 1977 by Robert M. Haft as one of the companies in the Dart Group, which was owned by the Haft family.[61] Crown Books began as a bookstore chain, then later added audio and video products in 1985.[62] One of the largest retail bookstore chains in the United States, Crown Books included more than 160 bookstores in seven metropolitan areas nationwide: Washington, D.C., Los Angeles, Chicago, San Francisco, San Diego, Houston, and Seattle.[63] It operated as a centralized organization headquartered in Maryland, USA. Crown Books knew the value of automation in support of decision-making. Stores were equipped with "computerized point-of-sale/inventory management systems, with standardized store interior layouts designed to enable rapid opening of new stores while maximizing efficiency" for the purpose of making economic use of store space.[64] This is one example of Crown Books' attention to what Gartner calls the core area of business "Automation to Support Remote Operations". Late in 1996, in an effort to make its vast recordkeeping requirements less costly and time-consuming, the company eliminated its original system and implemented a new, closed-loop information system capable of correlating operations and decision support functions.[65] Solutions to such problems as inventory shortages, pricing changes, order cancellations, and inter-store stock transfers could now be quickly accessed, along with immediate knowledge of any changes to bottom-line profits.[66] This is another example of Crown Books' actions in the core area of business known as "Automation to Support Remote Operations".

Most Crown Books stores were located in strip shopping centers and urban streets rather than in enclosed shopping malls, and were positioned in clusters to maximize advertising dollars and reduce distribution costs.[67] In 1995, the Securities and Exchange Commission (SEC) began investigating the Dart Group, which held a controlling interest in the Crown Books.[68] Part of the SEC investigation involved Crown Books' failure to disclose information regarding financial difficulties at one of the company's subsidiaries.[69] The investigation led to a drop in the price of company stock. In 1996 Crown was listed as co-plaintiff in a lawsuit brought against Herbert H. Haft by the Dart Group charging fraud and breach of fiduciary duty with regard to business transactions made during the course of Haft's divorce and resulting power struggle.[70] A standstill rider was entered in court restricting certain relevant actions of the Dart Corporation until such time as all legal matters were resolved. A conditional settlement had been reached with Herbert H. Haft whereby he would relinquish his position and voting rights in the Dart Group in exchange for a monetary settlement.[71]

In 1997, when Crown Books employed over 1,400 full-time and more than 1,800 part-time associates, the company began a college recruitment program to ensure that its sales staff were knowledgeable and up-to-date on popular literary trends.[72] Facing stiff competition from bookstores including Barnes & Noble, Borders, and Books-A-Million, Crown Books began closing some of its stores. Dart Group was acquired by Richfood Holdings Inc., in 1998, when Dart's non-supermarket subsidiaries, including Crown Books, were sold.[73] With the dissolution of the Dart Group, Crown Books was unable to find a buyer and was forced into bankruptcy in 1998.[74]

As an independent store, Crown Books faced two hurdles that proved insurmountable. The first was coming out of bankruptcy. The second was competition from other bookstores, in particular Amazon, which had started to sell books on the Internet. Crown Books, a centralized organization that functioned in line with a Somewhere Operations business model, yielded market share to Amazon, a distributed organization that functions according to the Anywhere Operations business model. The liquidation of Crown Books Corporation stores was completed in 2001.[75]

Crown Books was liquidated at the turn of the century, when Gartner had not yet published its trend in business operations. The bookstore industry and the available technology were not mature enough to formally practise all of the core areas of business that Gartner would years later identify as Collaboration and Productivity, Secure Remote Access, Cloud and Edge Infrastructure, Quantification of the Digital Experience, and Automation to Support Remote Operations.

Anywhere Operations – Amazon Online Bookstore

Gartner's expression "Anywhere Operations" refers to a contemporary business operating model built on a distributed infrastructure with supporting operations that accommodate stores in diverse geographic locations.[76] An Anywhere Operations business model benefits an organization by providing flexibility, and broadening the talent landscape, since Infrastructure and Operations teams are not tied to recruiting from a specific geographic location.[77] This shift from physical stores in Somewhere Operations to digital stores in Anywhere Operations forces IT executives to develop flexible and resilient organizations that enable staff to work from anywhere, allow customers to access services from everywhere, and foster the deployment of business services across distributed infrastructures.[78] The structured processes within traditional Infrastructure and Operations limit organizations when it comes to the flexibility of location. Anywhere Operations enables organizations to decentralize staff, activate operations, and make way for broader talent choices since organizations do not need to recruit staff in a specific geographic area. What is essential for Anywhere Operations is developing programmable infrastructure that enables the right work in the right place at the right time – this is the crux of optimal infrastructure.[79]

Due to the Coronavirus pandemic, more companies are allowing employees work from home. The trend falls under the theme of "location independence" which requires a technology shift to support customers and a new version of business.[80] With this operating model, a business can manage the deployment of services across distributed infrastructures while employees, customers, and business partners operate in physically remote environments.[81] Anywhere Operations is becoming the new default for business. It supports customers anywhere, hires employees from anywhere, and manages the deployment of business services across a distributed infractructure.[82] IT companies implement the Anywhere Operations business model by developing AI algorithms to automatically extract information from data based on predetermined criteria to enhance the manual process of decision-making.[83]

The path to automation in support of remote operations involves AI engineering, a discipline about governance and life cycle management of a wide range of operationalized AI and decision models. AI engineering has three major components: DataOps, ModelOps and DevOps.[84]

- **DataOps (Data Operations):** IBM defines DataOps as the orchestration of people, process, and technology to deliver reliable, high-quality data quickly.[85] The practice is focused on enabling collaboration across an organization to drive agility, speed, and new data initiatives at scale. Using the power of automation, DataOps is designed to solve challenges associated with inefficiencies in accessing, preparing, integrating and making data available.[86] DataOps supports team productivity with automation technology to deliver efficiency gains in project outputs and time. To experience the benefits of DataOps, the internal culture needs to evolve to truly be data-driven.[87]

- **ModelOps (Model Operations):** ModelOps is an approach to automation of operations that synchronizes application development activities with model pipelines for migration to production.[88] With ModelOps, an organization can optimize data management and AI investments using data, models and resources from edge to cloud. ModelOps covers the end-to-end lifecycles for optimizing the use of models and applications across cloud platforms, accomplished by automation involving Machine Learning models, optimization models and other operational models to combine with Continuous Integration and Continuous Deployment (CICD).[89]

- **DevOps (Development & Operations):** The expression "DevOps" is an abbreviation of development and operations. DevOps is an integration of people, process and technology to continually provide value to customers.[90] DevOps enables formerly siloed roles —- application developers, computer operators, quality engineers and security personnel —- to coordinate their activities in producing more reliable products. By adopting a DevOps culture, teams gain the ability to better respond to customer needs, increase confidence in the applications they build, and achieve business goals faster.[91] This improved collaboration is integral to achieving business goals like accelerating time to market, maintaining system stability and improving recovery time.[92] Where an organization has multiple stand-alone organizational units, DevOps can integrate the technologies from separate units through hyperautomation. Hyperautomation is a process in which organizations automate as many business and IT processes as possible using tools like Machine Learning, event-driven software, robotic process automation, and other types of decision process and task automation tools.[93] Hyperautomation enables organizations to automate processes and orchestrate automation across functional areas.[94]

Earlier in this chapter, I provided descriptions of Gartner's core areas of business: Collaborate and Productivity, Secure Remote Access, Cloud and Edge Infrastructure, Quantification of the Digital Experience and Automation to Support Remote Operations. Although Gartner's core areas of business are defined separately, they do not operate independently of each other; they build on and reinforce each other.[95] Together they enable organizational flexibility that guides organizations in Anywhere Operations.

I choose Amazon as an example of an organization that uses an Anywhere Operations business model.

Amazon Online Bookstore

Amazon Online Bookstore is the organization I select for illustrating Gartner's Anywhere Operations business model. News articles and company history indicate that Amazon applies all five of the core areas of business that Garner identifies as supportive of Anywhere Operations: Collaborate and Productivity, Secure Remote Access, Cloud and Edge Infrastructure, Quantification of the Digital Experience and Automation to Support Remote Operations.

Jeff Bezos founded Amazon as an online bookseller in 1995 when it offered the world's largest collection of books to customers who have Internet access.[96] Orders were processed online and delivered to customers' addresses all over the world.[97] Bezos did not regard Amazon simply as a retailer selling books online. In his view, Amazon was a technology company whose core business was the simplification of online transactions for customers.[98] Two years after being founded, Amazon had a customer base of roughly 1,000,000 registered accounts.[99] In 1998, Amazon began selling musical products and computer games, while expanding its services internationally by buying other online bookstores that were based in the United Kingdom and Germany.[100] By the beginning of the 21st century, Amazon expanded its range of products to include consumer electronics, video games, software, home-improvement items, toys and games.[101] The addition of Amazon Web Services (AWS), was consistent with the founder's intention of establishing Amazon as a technology company, not just an online retailer.[102]

During the first decade of the 21st century, Amazon was applying Gartner's core area of business known as Cloud Storage and Edge Computing. The evidence is noticeable in the fact that AWS took advantage of cloud computing. Amazon Mechanical Turk was using crowdsourcing, and Amazon Kindle began offering eBooks for sale. For fast performance, edge applications used cloud computing to support transaction processing, production of analytics, storage of data, and Machine Learning algorithms. In addition, edge applications do their processing near to where data are generated, in order to deliver efficient real-time responsiveness, and reduce the volume of data transferred.[103] AWS' edge computing services provide edge infrastructure and software that perform data processing and analysis as close to the end-point as possible. To deliver edge computing, Amazon deploys AWS managed hardware and software to locations outside AWS data centers, as well as to devices that are owned by customers.[104] These initiatives are examples of Amazon's use of the core area of business that Gartner labels Cloud and Edge Infrastructure.

Amazon uses available technology to improve its efficiency and customer service. Amazon uses Artificial Intelligence to process orders and to test opportunities for using drones and robots to perform order fulfillment and delivery.[105] This is an example of Amazon's implementation of the core area of business that Gartner calls Automation to Support Remote Operations. Another example of Automation to Support Remote Operations is Amazon's acquisition of antennas to enable cost effectiveness in transferring satellite data from space.[106]

When Coronavirus restrictions are lifted, Amazon's corporate and technical employees will not have to work in offices full time. The Seattle Times reports that Amazon will allow those workers to work remotely two days per week.[107] This applies to employees in the greater Seattle area, as well as employees who worked in downtown offices before the pandemic.[108] Working remotely allows employees to be productive by collaborating with each other and with customers when they are not physically in Amazon offices. This is an example of Amazon applying the core area of business that Gartner identifies as Collaboration and Productivity.

Amazon provides customers with a written guide for using AWS web services. The AWS guide includes instructions to set up secured server remote access sessions with multi-factor authentication (MFA) using AWS Session Manager and AWS Single Sign-On.[109] This architecture reduces risk to security associated with access to AWS. This is an example of Amazon applying the core area of business known as Secure Remote Access.

QuickSight is Amazon's business analytics service that enables AWS customers to produce dashboards of quantification about their business. QuickSight is a Business Intelligence (BI) service that offers customers features for creating and publishing interactive BI dashboards.[110] Since the service resides on the cloud, QuickSight dashboards can be accessed from any device, and embedded into applications, portals, and websites.[111] Being serverless, QuickSight can scale to thousands of users without any modifications in infrastructure management or capacity planning. This BI service allows customers to pay only when users access dashboards or reports, so it is cost-effective for large scale deployments.[112] Customers can pose business questions about data to QuickSight in ordinary language and get answers in seconds.[113] QuickSight is an example of the core area of business that Gartner describes as "Quantification to Support Remote Operations".

Hybrid Operations – Barnes & Noble Bookstore

"Hybrid Operations" is the expression I use to refer to a business model that is part "Somewhere Operations" and part "Anywhere Operations". I choose Barnes & Noble as an example of a Hybrid Operations, because it operates partly as a brick-and-mortar store, and partly as an online store. The transition from a traditional business model of Somewhere Operations to a contemporary business model of Anywhere Operations involves a cultural change, because it challenges the traditional thinking of providing infrastructure and supporting operations from one centralized location. Instead of a direct transition from Somewhere Operations to Anywhere Operations, an interim transition to Hybrid Operations may be less disruptive of productivity.[114] To initiate the transition, the Infrastructure & Operation leaders need to assess the feasibility for each business unit, to determine where remote working makes sense, and develop plans to enable teams to dynamically work both on-site and remotely.[115] In Hybrid Operations, employees treat their offices as business hubs while continuing to be productive from their home workspaces.[116] The hybrid approach does not eliminate the need for physical spaces, however, any use of physical space needs to be digitally enhanced and seamlessly delivered.[117]

The Hybrid Operations business model lends itself to ensuring a blend of virtual and physical experiences for an organization's employees and customers. This model presupposes an organization that has a mix of local operations and global operations. The local operations cover a shift per day availability of business services, when employees have in-store, in-person interaction with customers. The global operations covers round-the-clock support for customers, enablement for employees and availability for business services from anywhere in the world.[118] In a transition to Anywhere Operations, there is a need for a hybrid workspace in which employees use their office as a business hub and their home as a replica, where they can work remotely, and as efficiently.[119] The hybrid workforce and dependence of business on digital tools necessitate a transformation of the IT infrastructure to support cybersecurity solutions, and support remote services to enable the blend of virtual and real experiences.[120] Business needs and the Coronavirus pandemic have focused a spotlight on the facts that the boundaries between real and virtual environments have been blurring, also that organizations have to ensure uninterrupted access to corporate resources regardless of the place and time.[121]

Barnes & Noble Bookstore

Barnes & Noble's approach to the Hybrid Operations business model is not just one of transition from Somewhere Operations to Anywhere Operations. Barnes & Noble appears to intentionally sustain both business models; Somewhere Operations for its chain of brick-and-mortar stores, and Anywhere Operations for its online store.

Leonard Riggio acquired the Barnes & Noble trade name in 1971, after which he grew the brick-and-mortar store in New York City into "The World's Largest Bookstore" offering textbooks and trade titles.[122] In 1993, Barnes & Noble became a publicly traded company, later adding an e-commerce website and publishing capabilities. In 2009, the company entered the eBook market and launched its NOOK® brand of e-Reading products, which provide an immersive digital reading experience. Over the past several years, Barnes & Noble has introduced several devices in the tablet and e-Reader categories. [123]

Barnes & Noble's Chief Executive, James Daunt, is leaving behind the strategy that, for several decades, made it a major bookseller.[124] In the past, it was enough to maximize economies of scale and simplify the in-store shopping experience. Now that Barnes & Noble competes with Amazon, Daunt is empowering store managers to populate bookshelves based on the preferences of local customers. New York-based book buyers will no longer be the ones to choose which titles appear on shelves.[125] That is one way in which Barnes & Noble is addressing declining revenue due to Amazon's dominance in the market.[126] Giving store managers autonomy to make decisions based on local preferences, Barnes & Noble expects to give customers a shopping experience that is customized to their preferences.[127] This strategy of empowering individual store managers to satisfy local shoppers is an example of the core area of business that Gartner identifies as "Collaboration and Productivity".

In 2020, Barnes & Noble reported a data breach that impacted the e-book Nook services and possibly exposed customer data.[128] The breach began as a series of service disruptions, including customers not being able to access their digital libraries, login failures, and missing purchases. The disruptions extended to some physical Point-of-Sale devices.[129] Barnes & Noble confirmed that the disruption was caused by malware. In an e-mail to customers, the bookseller acknowledged a digital intrusion, leading to "unauthorized and unlawful access to certain Barnes & Noble corporate systems." [130] This hacking story indicates Barnes & Noble may not have adequate security for its e-book NOOK services. It appears that Barnes & Noble has a flaw in the core area of business that Gartner identifies as Secure Remote Access.

Barnes & Noble announced that it would embed Near Field Communication (NFC) chips in the bookseller's NOOK devices. That was intended to attract more in-store visits by its Nook e-Readers. Barnes & Noble CEO William Lynch made this announcement during an interview with Fortune magazine in 2012:[131]

> "We can work with the publishers so they would ship a copy of each hardcover with an NFC chip embedded with all the editorial reviews they can get on BN.com," Lynch said in the interview. "And if you had your Nook, you can walk up to any of our pictures, any of our aisles, any of our bestseller lists, and just touch the book, and get information on that physical book on your Nook and have some frictionless purchase experience."

A shopper can take their NOOK device into any store, use it to touch a physical book, find out information about the book, and purchase the book, all in one seamless shopping experience. The Nook also features a Read in Store capability that allows visitors to stream and read any book for up to one hour while shopping in a Barnes & Noble bookstore.[132] Integrating NFC technology into the Nook enables a diversity of applications intended to provide customers with a more engaging, interactive shopping experience that includes information access and mobile marketing. For example, Barnes & Noble displays an NFC-tagged copy of each best-seller in its stores, so that customers can tap the tags to be directed automatically to online reviews and other information related to that title. Essentially, this initiative brings together the best aspects of online and offline shopping—that is, the ability to access reviews and product information with the potential to touch (and purchase) the item immediately—at brick-and-mortar locations.[133] The use of NFC technology to enable access to product information stored in the cloud, and activate edge applications for data processing in stores at the point when customers touch their NOOK devices against physical books, indicate that Barnes & Noble is applying the core area of business that Gartner identifies as Cloud and Edge Infrastructure.

Since most of its customer outreach is conducted online, via e-mail newsletters, Barnes & Noble wanted to personalize and improve its digital marketing process, in order to strengthen its customer relationships. It was looking for a reliable software development partner to help achieve this objective, so it could focus its resources more on its core business. After evaluating several vendors, Barnes & Noble chose Folio3 as its development partner, due to Folio3's extensive expertise in workflow automation services, web-based applications and its reputation for delivering high quality solutions.[134] The solution is based on a customized Content Management System (CMS) that allows Barnes & Noble's marketing team to rapidly create e-mail newsletters using pre-defined templates. The newsletters inform customers about new releases, pre-orders, bestsellers, NOOK daily finds, and other offers or recommendations. The CMS is part of an automated workflow that enables the marketing team to collaborate and finalize the content and structure of each newsletter.[135] Once the newsletter is created, the system allows the marketing team to create the target list of customers for each e-mail blast. The system also enables the marketing team to modify the list of recipients at any time prior to sending out the e-mails. Due to Folio3's workflow automation services, Barnes & Noble's marketing team is able to send out highly customized newsletters to millions of customers at any time. Contracting with Folio3 to develop workflow automation services for interaction with customers who are remotely located, indicates that Barnes & Noble is applying the core area of business that Gartner calls Automation to Support Remote Operations.

Comparing Gartner's Trend In Business Operations With Goodhart's Trend In Demography

Goodhart makes note of a qualitative difference between Somewheres and Anywheres. The Somewheres are people who have ascribed identities, while the Anywheres have achieved identities. The Somewheres inherit their identity based on characteristics that are not of their own making, while the Anywheres construct their identity from characteristics that are of their own making. Here I compare Gartner's trend in business operations with Goodhart's trend in demography. Goodhart's trend was described in detail in Chapter 3: "David Goodhart's Trend In Demography".

As a reminder, here is a summary of Goodhart's trend.

There is a cleavage of Western populations into …

A population of "Somewheres" who define their identity
by their geographic stability
and their heritage, and

A population of "Anywheres" who define their identity
by their geographic mobility
and their accomplishments.

Gartner brings our attention to a trend in business operations, while Goodhart points out a trend in demography. Gartner's (Somewhere) Operations business model is rooted geographically, and it relies on in-person interaction with customers during specific office hours. Goodhart's Somewheres are rooted to specific geographic locations with ties to local, in-person community. What they have in common are specificity of geographic location, reliance on in-person communication, and availability for interaction during specific time zones. I interpret this to mean that Gartner's (Somewhere) Operations are populated by people who fall into the group Goodhart calls Somewheres.

Gartner's Anywhere Operations business model is geographically dispersed, engages customers in digital communication, and operates around the clock without specific hours. Goodhart's Anywheres are geographically mobile with ties to global networks for remote interaction with others. What they have in common are geographic mobility, and the flexibility of digital communication with others, and unlimited office hours. What I interpret here is that Gartner's Anywhere Operations are populated by people Goodhart labels Anywheres.

I see the Anywheres not merely as a geographically mobile generation emerging from the geographically rooted Somewheres, but also as being cognitively and psychologically different from the Somewheres. Cognitively, the Somewheres acquire knowledge from traditional means of education, in which people learn subjects that they intentionally select. Their knowledge acquisition is aimed at explicit knowledge. Anywheres also acquire knowledge in that way, but in addition, they acquire knowledge that they do not intentionally select. Their knowledge acquisition includes implicit knowledge. The Anywheres are able to acquire implicit knowledge because technology offers Machine Learning which is capable of extracting both explicit and implicit knowledge by mining large collections of data.

Gartner's distinction between (Somewhere) Operations and Anywhere Operations is not merely about organizations making a transition from brick-and-mortar stores to online stores. It is also about people changing the way they think. From cognition that is built from the addition of traditional, incremental knowledge, to cognition that is re-constructed in contemporary re-organization of knowledge. From a mindset of centralization to one of decentralization. From communicating with people we see and know to digital communication with people we may not see or know. From functioning in the context of local cultural norms to functioning in the context of global cultural norms.

Since David Goodhart's Anywheres are comfortable with changes brought about by technology, they welcome the many modalities of

Gartner's multiexperience in support of their mobile lifestyles across varied geographic locations. By being receptive to technology and willing to be geographically mobile, the Anywheres are poised to take advantage of new developments in Anywhere Operations. This type of operation requires a mindset that is more cognitively flexible than Somewhere Operations. The ability to embrace the core areas of business that Gartner identifies sets the Anywheres ahead in terms of survival in the context of evolving business models.

Next, I compare Gartner's trend in business operations with Giegerich's trend in psychology.

Comparing Gartner's Trend In Business Operations With Giegerich's Trend In Psychology

Giegerich's trend was described in detail in Chapter 2: "Wolfgang Giegerich's Trend In Psychology".

Here is a reminder of my summary of Giegerich's trend.

There is a paradigm shift ...

From a focus on the semantical level of psychology,
where individuals engage in the individuation process,
a goal-seeking effort to differentiate their minds from the
unconsciousness of their communities,

To a focus on the syntactical level of psychology,
where human culture engages in the interiorization process,
an intellectual discipline of interpreting phenomena
that emerge in the world

At the semantical level of psychology, consciousness grows by expansion of content and meaning. At the syntactical level of psychology, consciousness grows by restructuring of content and outlook. When individuals define themselves in terms of their families and organizations define themselves in terms of their local geography, consciousness grows by incremental addition of content, building block style. When people learn existing social norms, they are expanding their consciousness by acquiring knowledge of traditions and cultures already established. When people define themselves in terms of their mobility and achievements and organizations define themselves in terms of global geography and digital networks, consciousness grows by restructuring of content and outlook.

When people acquire traditional norms by remaining geographically rooted and remaining in established communities, they sustain existing mental constructs of identity, time and geography. When people create new norms by becoming mobile and joining digital networks, they restructure their mental constructs of identity, time and geography. By implication, they are also restructuring their consciousness so they can function in circumstances different from the cultural norms that exist in the circumstances where they grew up.

The semantical and syntactical levels of psychology can be seen in Somewhere Operations and Anywhere Operations. Gartner describes significant differences between (Somewhere) Operations and Anywhere Operations: they entail different mental constructs, and different levels of psychology. These are the reasons I propose that Somewhere Operations and Anywhere Operations involve people who are psychologically different. Their sense of identity, their mental constructs of time and space, their aptitude for sense-making are all different. Somewheres have a sense of identity that is inherited from the family in which they were born and the community in which they grew up, while Anywheres have a sense of identity that they constructed to suit their mobile lifestyle. The Somewheres inherit their identity based on characteristics that are not of their own making, while the Anywheres construct their identity from characteristics that are of their own making.

Psychologically, the Somewheres inherit their identity, while the Anywheres construct their identity. That involves restructuring of consciousness. Somewheres have a mental construct of time that is attuned to a specific time zone, while Anywheres have a mental construct of time that is attuned to multiple time zones around the globe. Somewheres have a mental construct of space that is geographically local, while Anywheres have a mental construct of space that is geographically global. Somewheres have a sense of geographic rootedness that places identity in stability and familiarity. Anywheres have a sense of geographic mobility that requires the ability to carve out a path in life, to navigate that path with an assured sense of identity, and respond with resilience when there are changes in the environment. Flexibility in sense of identity, sense of time, sense of space, and the ability to navigate new paths in life all point to the Anywheres functioning at the syntactical level of psychology, where the ongoing acquisition of new knowledge accompanied by changing outlooks necessitate the restructuring of consciousness.

Somewheres have an aptitude for sense-making that involves explicit knowledge derived from human learning, while Anywheres have an aptitude for sense-making that involves both explicit and implicit knowledge. Implicit knowledge comes from Machine Learning, in which algorithms extract explicit and implicit knowledge by mining large volumes of data. Somewheres have characteristics that are compatible with the Somewhere Operations business model, where there are few changing variables, while Anywheres have characteristics that are compatible with the Anywhere Operations business model, where there are multiple changing variables. For example, Somewheres contend with one reality, while Anywheres contend with multiple realities: Augmented Reality, Virtual Reality, and Mixed Reality.

My interpretation is that the Somewheres function at Giegerich's semantical level of psychology. Their consciousness is grounded in a sense of identity that is based on what they inherited from family and local community. Their consciousness expands incrementally by a sense-making aptitude that is derived from traditional human learning by which they accumulate explicit knowledge. Their single-time-zone sense of time and their local-geography sense of space, guided by their national culture, all place limits on the meaning they can construct from their experience. Together, these characteristics point to a consciousness that grows by expansion of content. These characteristics indicate that the Somewheres function primarily at the semantical level of psychology.

In my opinion, the Anywheres function at Giegerich's syntactical level of psychology. Their consciousness is grounded in a sense of identity that is based on their own achievements, and is restructured when there are new achievements. Their consciousness is restructured periodically by a sense-making aptitude that is derived from a combination of human learning and Machine Learning, by which they acquire a mix of explicit and implicit knowledge. Their sense of time is a mental river whose mainstream and tributaries enable the structuring of experience. Their sense of space is a mental scaffolding for constructing meaning from their experience, guided by the variables in a global culture. Together, these characteristics point to a consciousness that grows by restructuring of outlook. These characteristics indicate that the Anywheres function primarily at the syntactical level of psychology.

The trends discerned by Gartner, Goodhart and Giegerich support my view that bifurcations in multiple disciplines contribute to an overarching paradigm shift that puts Homo Sapiens on the brink of generating a new version of humanity.

Summary

This chapter shows some correspondence among David Goodhart's trend in demography, Gartner's trend in business operations and Wolfgang Giegerich's trend in psychology. All three trends are about a new generation emerging from a traditional generation with a qualitatively different outlook. All three trends have a focus on collective groups. All the trends have been progressing actively in the early decades of the 21st century. All are progressing to a higher level of human functionality. In all the trends, technology plays a strong role of supporting collective groups by becoming increasingly people-literate.

Gartner's trend in business operations indicates a bifurcation into physical business operations and digital business operations. In the previous chapter, Goodhart's trend in demography depicted a bifurcation of identity into ascribed identity and achieved identity. In an earlier chapter, Giegerich's trend in psychology portrayed a bifurcation of human sense of agency into a sense of agency that dwells in individuals, and a sense of agency that resides in human culture as a whole.

Business operations, demography and psychology embody dynamical systems that have variables whose patterns of behavior change over time. When variables in dynamical systems bifurcate, the systems become unstable and predisposed to change. When the magnitude of the bifurcations is large, it can bring about transformation in the systems. The bifurcations in business operations, demography and psychology create the possibility of major changes for Homo Sapiens species. Combined, the bifurcations occurring in business operations, demography and psychology support my proposal that Homo Sapiens on the cusp of generating a new version of humanity.

NOTES:

1. See Gartner's web site: https://www.gartner.com/smarterwith-gartner/gartner-top-strategic-technology-trends-for-2021/, Article "Gartner Top Strategic Technology Trends for 2021" by Kasey Panetta.
2. Gartner's web site: www.Gartner.com.
3. Forbes' web site: www.Forbes.com.
4. Stefanini GROUP's web site: www.stefanini.com.
5. IBM's web site: www.IBM.com.

6. *"The Road to Somewhere: The Populist Revolt and the Future of Politics"* by demographer David Goodhart, C. Hurst & Co. (Publishers) Ltd, 2017.
7. Amazon' web site: www.amazon.com.
8. Barnes & Noble's web site: www.bn.com.
9. Crown Publishing Group's web site: www.crownpublishing.com.
10. See Gartner's web site: https://www.gartner.com/smarterwith-gartner/gartner-top-strategic-technology-trends-for-2021/, Article "Gartner Top Strategic Technology Trends for 2021" by Kasey Panetta.
11. See web site: Security InfoWatch, https://www.securityinfowatch.com/cybersecurity/press-release/21159433/gartner-gartner-identi-fies-the-top-strategic-technology-trends-for-2021, Article "Gartner Identifies the Top Strategic Technology Trends for 2021".
12. See web site: Stefanini GROUP, 4 Ways to Successfully Support Gartner Tech Trend Anywhere Operations | Stefanini, Article "4 Ways to Successfully Apply Gartner Tech Trend 'Anywhere Operations' ".
13. See web site: Forbes, Anywhere Operations Model: How IT Companies Implemented Them First (forbes.com), Article "Anywhere Operations Model: How IT Companies Implemented Them First" by Ilya Gandzeichuk, Forbes Technology Council Member.
14. See web site: Forbes, Anywhere Operations Model: How IT Companies Implemented Them First (forbes.com), Article "Anywhere Operations Model: How IT Companies Implemented Them First" by Ilya Gandzeichuk, Forbes Technology Council Member.
15. See web site: Forbes, Anywhere Operations Model: How IT Companies Implemented Them First (forbes.com), Article "Anywhere Operations Model: How IT Companies Implemented Them First" by Ilya Gandzeichuk, Forbes Technology Council Member.

16. See web site: Forbes, <u>Anywhere Operations Model: How IT Companies Implemented Them First (forbes.com)</u>, Article "Anywhere Operations Model: How IT Companies Implemented Them First" by Ilya Gandzeichuk, Forbes Technology Council Member.

17. See web site: Forbes, <u>Anywhere Operations Model: How IT Companies Implemented Them First (forbes.com)</u>, Article "Anywhere Operations Model: How IT Companies Implemented Them First" by Ilya Gandzeichuk, Forbes Technology Council Member.

18. See web site: Forbes, <u>Anywhere Operations Model: How IT Companies Implemented Them First (forbes.com)</u>, Article "Anywhere Operations Model: How IT Companies Implemented Them First" by Ilya Gandzeichuk, Forbes Technology Council Member.

19. See web site: CMSWire, <u>Why Your Digital Workplace Needs a CCP (cmswire.com)</u>, Article "Why Your Digital Workplace Needs a CCP" by David Roe.

20. See web site: CMSWire, <u>Why Your Digital Workplace Needs a CCP (cmswire.com)</u>, Article "Why Your Digital Workplace Needs a CCP" by David Roe.

21. See web site: CMSWire, <u>Why Your Digital Workplace Needs a CCP (cmswire.com)</u>, Article "Why Your Digital Workplace Needs a CCP" by David Roe.

22. See web site: Stefanini GROUP, <u>4 Ways to Successfully Support Gartner Tech Trend Anywhere Operations | Stefanini</u>, Article "4 Ways to Successfully Apply Gartner Tech Trend 'Anywhere Operations' ".

23. See web site: Stefanini GROUP, <u>4 Ways to Successfully Support Gartner Tech Trend Anywhere Operations | Stefanini</u>, Article "4 Ways to Successfully Apply Gartner Tech Trend 'Anywhere Operations' ".

24. See web site: Forbes, <u>Anywhere Operations Model: How IT Companies Implemented Them First (forbes.com)</u>, Article "Anywhere Operations Model: How IT Companies Implemented Them First" by Ilya Gandzeichuk, Forbes Technology Council Member.

25. See web site: Forbes, <u>Anywhere Operations Model: How IT Companies Implemented Them First (forbes.com)</u>, Article "Anywhere Operations Model: How IT Companies Implemented Them First" by Ilya Gandzeichuk, Forbes Technology Council Member.

26. See web site: techutzpah, <u>What is Cybersecurity Mesh? - techutzpah</u>, Article "What is Cybersecurity Mesh?" by Rajashree Rao.

27. See web site: techutzpah, <u>What is Cybersecurity Mesh? - techutzpah</u>, Article "What is Cybersecurity Mesh?" by Rajashree Rao.

28. See web site: techutzpah, <u>What is Cybersecurity Mesh? - techutzpah</u>, Article "What is Cybersecurity Mesh?" by Rajashree Rao.

29. See web site: techutzpah, <u>What is Cybersecurity Mesh? - techutzpah</u>, Article "What is Cybersecurity Mesh?" by Rajashree Rao.

30. See web site: techutzpah, <u>What is Cybersecurity Mesh? - techutzpah</u>, Article "What is Cybersecurity Mesh?" by Rajashree Rao.

31. See web site: techutzpah, <u>What is Cybersecurity Mesh? - techutzpah</u>, Article "What is Cybersecurity Mesh?" by Rajashree Rao.

32. See web site: techutzpah, <u>What is Cybersecurity Mesh? - techutzpah</u>, Article "What is Cybersecurity Mesh?" by Rajashree Rao.

33. See web site: techutzpah, <u>What is Cybersecurity Mesh? - techutzpah</u>, Article "What is Cybersecurity Mesh?" by Rajashree Rao.

34. See web site: MarketScreener, <u>Digital is the New Default – Gartner's Report on Top Strategic Technology Trends for 2021: Anywhere Operations | MarketScreener</u>, Article "Digital is the New Default – Gartner's Report on Top Strategic Technology Trends for 2021: Anywhere Operations".

35. See web site: Microsoft | Support, <u>What is: Multifactor Authentication (microsoft.com)</u>, Article "What is: Multifactor Authentication".

36. See web site: Microsoft | Support, <u>What is: Multifactor Authentication (microsoft.com)</u>, Article "What is: Multifactor Authentication".
37. See web site: Microsoft | Support, <u>What is: Multifactor Authentication (microsoft.com)</u>, Article "What is: Multifactor Authentication".
38. See web site: Forbes, <u>Anywhere Operations Model: How IT Companies Implemented Them First (forbes.com)</u>, Article "Anywhere Operations Model: How IT Companies Implemented Them First" by Ilya Gandzeichuk,_Forbes Technology Council Member.
39. See web site: Forbes, <u>Anywhere Operations Model: How IT Companies Implemented Them First (forbes.com)</u>, Article "Anywhere Operations Model: How IT Companies Implemented Them First" by Ilya Gandzeichuk,_Forbes Technology Council Member.
40. See web site: Forbes, <u>Anywhere Operations Model: How IT Companies Implemented Them First (forbes.com)</u>, Article "Anywhere Operations Model: How IT Companies Implemented Them First" by Ilya Gandzeichuk,_Forbes Technology Council Member.
41. See web site: Smarter With Gartner, <u>What Edge Computing Means for Infrastructure and Operations Leaders - Smarter With Gartner</u>, Article "What Edge Computing Means for Infrastructure and Operations Leaders".
42. See web site: Smarter With Gartner, <u>What Edge Computing Means for Infrastructure and Operations Leaders - Smarter With Gartner</u>, Article "What Edge Computing Means for Infrastructure and Operations Leaders".
43. See web site: Smarter With Gartner, <u>What Edge Computing Means for Infrastructure and Operations Leaders - Smarter With Gartner</u>, Article "What Edge Computing Means for Infrastructure and Operations Leaders".

44. See web site: Smarter With Gartner, <u>What Edge Computing Means for Infrastructure and Operations Leaders - Smarter With Gartner</u>, Article "What Edge Computing Means for Infrastructure and Operations Leaders".

45. See web site: Smarter With Gartner, <u>What Edge Computing Means for Infrastructure and Operations Leaders - Smarter With Gartner</u>, Article "What Edge Computing Means for Infrastructure and Operations Leaders".

46. See web site: CDO Trends, <u>Gartner Predicts the Future of Cloud and Edge Infrastructure | CDOTrends</u>, Article "Gartner Predicts the Future of Cloud and Edge Infrastructure" by John McArthur.

47. See web site: Smarter With Gartner, <u>What Edge Computing Means for Infrastructure and Operations Leaders - Smarter With Gartner</u>, Article "What Edge Computing Means for Infrastructure and Operations Leaders".

48. See web site: Smarter With Gartner, <u>What Edge Computing Means for Infrastructure and Operations Leaders - Smarter With Gartner</u>, Article "What Edge Computing Means for Infrastructure and Operations Leaders".

49. See web site: Smarter With Gartner, <u>What Edge Computing Means for Infrastructure and Operations Leaders - Smarter With Gartner</u>, Article "What Edge Computing Means for Infrastructure and Operations Leaders".

50. See web site: CDO Trends, <u>Gartner Predicts the Future of Cloud and Edge Infrastructure | CDOTrends</u>, Article "Gartner Predicts the Future of Cloud and Edge Infrastructure" by John McArthur.

51. See web site: CDO Trends, <u>Gartner Predicts the Future of Cloud and Edge Infrastructure | CDOTrends</u>, Article "Gartner Predicts the Future of Cloud and Edge Infrastructure" by John McArthur.

52. See web site: MarketScreener, <u>Digital is the New Default – Gartner's Report on Top Strategic Technology Trends for 2021: Anywhere Operations | MarketScreener</u>, Article "Digital is the New Default – Gartner's Report on Top Strategic Technology Trends for 2021: Anywhere Operations".

53. See web site: MarketScreener, <u>Digital is the New Default –</u>
<u>Gartner's Report on Top Strategic Technology Trends for 2021:</u>
<u>Anywhere Operations | MarketScreener</u>, Article "Digital is the
New Default – Gartner's Report on Top Strategic Technology
Trends for 2021: Anywhere Operations".
54. See web site: MarketScreener, <u>Digital is the New Default –</u>
<u>Gartner's Report on Top Strategic Technology Trends for 2021:</u>
<u>Anywhere Operations | MarketScreener</u>, Article "Digital is the
New Default – Gartner's Report on Top Strategic Technology
Trends for 2021: Anywhere Operations".
55. See web site: MarketScreener, <u>Digital is the New Default –</u>
<u>Gartner's Report on Top Strategic Technology Trends for 2021:</u>
<u>Anywhere Operations | MarketScreener</u>, Article "Digital is the
New Default – Gartner's Report on Top Strategic Technology
Trends for 2021: Anywhere Operations".
56. See web site: MarketScreener, <u>Digital is the New Default –</u>
<u>Gartner's Report on Top Strategic Technology Trends for 2021:</u>
<u>Anywhere Operations | MarketScreener</u>, Article "Digital is the
New Default – Gartner's Report on Top Strategic Technology
Trends for 2021: Anywhere Operations".
57. See web site: AI Multiple, <u>AIOps: Guide to integrating AI into IT</u>
<u>Operations in 2021 (aimultiple.com)</u>, Article "AIOps: Guide to
integrating AI into IT Operations in 2021".
58. See web site: AI Multiple, <u>AIOps: Guide to integrating AI into IT</u>
<u>Operations in 2021 (aimultiple.com)</u>, Article "AIOps: Guide to
integrating AI into IT Operations in 2021".
59. See web site: Stefanini GROUP, <u>4 Ways to Successfully Support</u>
<u>Gartner Tech Trend Anywhere Operations | Stefanini,</u> Article
"4 Ways to Successfully Apply Gartner Tech Trend 'Anywhere
Operations' ".
60. See web site: Stefanini GROUP, <u>4 Ways to Successfully Support</u>
<u>Gartner Tech Trend Anywhere Operations | Stefanini,</u> Article
"4 Ways to Successfully Apply Gartner Tech Trend 'Anywhere
Operations' ".

61. See web site: Company-Histories.com, <u>Crown Books Corporation — Company History (company-histories.com)</u>, Article "Crown Books Corporation".

62. See web site: Company-Histories.com, <u>Crown Books Corporation — Company History (company-histories.com)</u>, Article "Crown Books Corporation".

63. See web site: Company-Histories.com, <u>Crown Books Corporation — Company History (company-histories.com)</u>, Article "Crown Books Corporation".

64. See web site: Company-Histories.com, <u>Crown Books Corporation — Company History (company-histories.com)</u>, Article "Crown Books Corporation".

65. See web site: Company-Histories.com, <u>Crown Books Corporation — Company History (company-histories.com)</u>, Article "Crown Books Corporation".

66. See web site: Company-Histories.com, <u>Crown Books Corporation — Company History (company-histories.com)</u>, Article "Crown Books Corporation".

67. See web site: Company-Histories.com, <u>Crown Books Corporation — Company History (company-histories.com)</u>, Article "Crown Books Corporation".

68. See web site: Company-Histories.com, <u>Crown Books Corporation — Company History (company-histories.com)</u>, Article "Crown Books Corporation".

69. See web site: Company-Histories.com, <u>Crown Books Corporation — Company History (company-histories.com)</u>, Article "Crown Books Corporation".

70. See web site: Company-Histories.com, <u>Crown Books Corporation — Company History (company-histories.com)</u>, Article "Crown Books Corporation".

71. See web site: Company-Histories.com, <u>Crown Books Corporation — Company History (company-histories.com)</u>, Article "Crown Books Corporation".

72. See web site: Company-Histories.com, <u>Crown Books Corporation — Company History (company-histories.com)</u>, Article "Crown Books Corporation".

73. See web site: The Baltimore Sun, <u>Crown hopes for new reign; It expects to leave bankruptcy, change format of stores; Booksellers - Baltimore Sun</u>, Article "Crown hopes for new reign; It expects to leave bankruptcy, change format of stores; Booksellers".

74. See web site: The Baltimore Sun, <u>Crown hopes for new reign; It expects to leave bankruptcy, change format of stores; Booksellers - Baltimore Sun</u>, Article "Crown hopes for new reign; It expects to leave bankruptcy, change format of stores; Booksellers".

75. See web site: Company-Histories.com, <u>Crown Books Corporation — C ompany History (company-histories.com)</u>, Article "Crown Books Corporation".

76. See web site: Gartner, <u>Infrastructure and Operations: Gartner Top 6 Trends for 2021</u>, Article "Gartner Top 6 Trends Impacting Infrastructure & Operations in 2021" by Megan Rimol.

77. See web site: Gartner, <u>Infrastructure and Operations: Gartner Top 6 Trends for 2021</u>, Article "Gartner Top 6 Trends Impacting Infrastructure & Operations in 2021" by Megan Rimol.

78. See web site: CHANNELe2e, <u>Gartner: 6 Top Infrastructure and Operations Trends to Watch in 2021 - ChannelE2E</u>, Article "Gartner: 6 Top Infrastructure and Operation Trends to Watch in 2012".

79. See web site: CHANNELe2e, <u>Gartner: 6 Top Infrastructure and Operations Trends to Watch in 2021 - ChannelE2E</u>, Article "Gartner: 6 Top Infrastructure and Operation Trends to Watch in 2012".

80. See web site: Stefanini GROUP, <u>4 Ways to Successfully Support Gartner Tech Trend Anywhere Operations | Stefanini</u>, Article "4 Ways to Successfully Apply Gartner Tech Trend 'Anywhere Operations' ".

81. See web site: Stefanini GROUP, 4 Ways to Successfully Support Gartner Tech Trend Anywhere Operations | Stefanini, Article "4 Ways to Successfully Apply Gartner Tech Trend 'Anywhere Operations' ".

82. See web site: Security InfoWatch, https://www.securityinfowatch.com/cybersecurity/press-release/21159433/gartner-gartner-identifies-the-top-strategic-technology-trends-for-2021, Article "Gartner identifies the top strategic trends for 2021".

83. See web site: Forbes, Anywhere Operations Model: How IT Companies Implemented Them First (forbes.com), Article "Anywhere Operations Model: How IT Companies Implemented Them First" by Ilya Gandzeichuk, Forbes Technology Council Member.

84. See web site: Gartner Newsroom, Gartner Identifies the Top Strategic Technology Trends for 2021, Article "Gartner Identifies the Top Strategic Technology Trends for 2021".

85. See web site: IBM.com, What is DataOps? - Journey to AI Blog (ibm.com), Article "What Is DataOps?".

86. See web site: IBM.com, What is DataOps? - Journey to AI Blog (ibm.com), Article "What Is DataOps?".

87. See web site: IBM.com, What is DataOps? - Journey to AI Blog (ibm.com), Article "What Is DataOps?".

88. See web site: IBM.com, IBM Watson Studio - IBM Data Science for ModelOps | IBM, Article "IBM Data Science for ModelOps".

89. See web site: IBM.com, IBM Watson Studio - IBM Data Science for ModelOps | IBM, Article "IBM Data Science for ModelOps".

90. See web site: Azure, What is DevOps? DevOps Explained | Microsoft Azure, Article "What Is DevOps?".

91. See web site: Azure, What is DevOps? DevOps Explained | Microsoft Azure, Article "What Is DevOps?".

92. See web site: Azure, What is DevOps? DevOps Explained | Microsoft Azure, Article "What Is DevOps?".

93. See web site: Gartner, top-tech-trends-ebook-2021.pdf (converge.com), Article "Top Strategic Technology Trends for 2021", edited by Brian Burke, Research Vice President, Gartner.

94. See web site: Gartner, top-tech-trends-ebook-2021.pdf (converge.com), Article "Top Strategic Technology Trends for 2021", edited by Brian Burke, Research Vice President, Gartner.

95. See web site: Gartner, top-tech-trends-ebook-2021.pdf (converge.com), Article "Top Strategic Technology Trends for 2021".

96. See web site: INTERESTING ENGINEERING, A Very Brief History of Amazon: The Everything Store | IE (interestingengineering.com), Article "A Very Brief History of Amazon: The Everything Store" by Christopher McFadden.

97. See web site: INTERESTING ENGINEERING, A Very Brief History of Amazon: The Everything Store | IE (interestingengineering.com), Article "A Very Brief History of Amazon: The Everything Store" by Christopher McFadden.

98. See web site: INTERESTING ENGINEERING, A Very Brief History of Amazon: The Everything Store | IE (interestingengineering.com), Article "A Very Brief History of Amazon: The Everything Store" by Christopher McFadden.

99. See web site: INTERESTING ENGINEERING, A Very Brief History of Amazon: The Everything Store | IE (interestingengineering.com), Article "A Very Brief History of Amazon: The Everything Store" by Christopher McFadden.

100. See web site: INTERESTING ENGINEERING, A Very Brief History of Amazon: The Everything Store | IE (interestingengineering.com), Article "A Very Brief History of Amazon: The Everything Store" by Christopher McFadden.

101. See web site: INTERESTING ENGINEERING, A Very Brief History of Amazon: The Everything Store | IE (interestingengineering.com), Article "A Very Brief History of Amazon: The Everything Store" by Christopher McFadden.

102. See web site: INTERESTING ENGINEERING, A Very Brief History of Amazon: The Everything Store | IE (interestingengineering.com), Article "A Very Brief History of Amazon: The Everything Store" by Christopher McFadden.

103. See web site: AWS, AWS for the Edge - Overview | Amazon Web Services, Article "ASW for the Edge".

104. See web site: AWS, <u>AWS for the Edge - Overview | Amazon Web Services</u>, Article "ASW for the Edge".

105. See web site: INTERESTING ENGINEERING, <u>A Very Brief History of Amazon: The Everything Store | IE (interestingengineering.com)</u>, Article "A Very Brief History of Amazon: The Everything Store" by Christopher McFadden.

106. See web site: INTERESTING ENGINEERING, <u>A Very Brief History of Amazon: The Everything Store | IE (interestingengineering.com)</u>, Article "A Very Brief History of Amazon: The Everything Store" by Christopher McFadden.

107. See web site: MarketWatch, <u>Amazon now says employees will be able to work from home 2 days a week - MarketWatch</u>, Article "Amazon now says employees will be able to work from home 2 days a week".

108. See web site: MarketWatch, <u>Amazon now says employees will be able to work from home 2 days a week - MarketWatch</u>, Article "Amazon now says employees will be able to work from home 2 days a week".

109. See PDF on Internet, <u>Securing Remote Access with Multi-Factor Authentication (awsstatic.com)</u>, Article " Securing Remote Access with Multi-Factor Authentication: Using AWS Systems Manager Session Manager and AWS Single Sign-On (AWS SSO)".

110. See web site: AWS, <u>Amazon QuickSight - Business Intelligence Service - Amazon Web Services</u>, Article "Amazon QuickSight".

111. See web site: AWS, <u>Amazon QuickSight - Business Intelligence Service - Amazon Web Services</u>, Article "Amazon QuickSight".

112. See web site: AWS, <u>Amazon QuickSight - Business Intelligence Service - Amazon Web Services</u>, Article "Amazon QuickSight".

113. See web site: AWS, <u>Amazon QuickSight - Business Intelligence Service - Amazon Web Services</u>, Article "Amazon QuickSight".

114. See web site: Stefanini GROUP, <u>4 Ways to Successfully Support Gartner Tech Trend Anywhere Operations | Stefanini</u>, Article "4 Ways to Successfully Apply Gartner Tech Trend 'Anywhere Operations' ".

115. See web site: Stefanini GROUP, <u>4 Ways to Successfully Support Gartner Tech Trend Anywhere Operations | Stefanini</u>, Article "4 Ways to Successfully Apply Gartner Tech Trend 'Anywhere Operations' ".

116. See web site: Gartner.com, <u>Infrastructure and Operations: Gartner Top 6 Trends for 2021</u>, Article "Smarter With Gartner".

117. See web site: Stefanini GROUP, <u>4 Ways to Successfully Support Gartner Tech Trend Anywhere Operations | Stefanini</u>, Article "4 Ways to Successfully Apply Gartner Tech Trend 'Anywhere Operations' ".

118. See web site: Forbes, <u>Anywhere Operations Model: How IT Companies Implemented Them First (forbes.com)</u>, Article "Anywhere Operations Model: How IT Companies Implemented Them First" by Ilya Gandzeichuk, Forbes Technology Council Member.

119. See web site: Forbes, <u>Anywhere Operations Model: How IT Companies Implemented Them First (forbes.com)</u>, Article "Anywhere Operations Model: How IT Companies Implemented Them First" by Ilya Gandzeichuk, Forbes Technology Council Member.

120. See web site: Forbes, <u>Anywhere Operations Model: How IT Companies Implemented Them First (forbes.com)</u>, Article "Anywhere Operations Model: How IT Companies Implemented Them First" by Ilya Gandzeichuk, Forbes Technology Council Member.

121. See web site: Forbes, <u>Anywhere Operations Model: How IT Companies Implemented Them First (forbes.com)</u>, Article "Anywhere Operations Model: How IT Companies Implemented Them First" by Ilya Gandzeichuk, Forbes Technology Council Member.

122. See web site: BARNES&NOBLE, <u>Barnes & Noble History (barnesandnobleinc.com)</u>, Article "History".

123. See web site: BARNES&NOBLE, <u>Barnes & Noble History (barnesandnobleinc.com)</u>, Article "History".

124. See web site: Forbes, <u>Will Barnes & Noble's Next Chapter Be Its Last? (forbes.com)</u>, Article "Will Barnes & Noble's Next Chapter Be Its Last?".

125. See web site: Forbes, <u>Will Barnes & Noble's Next Chapter Be Its Last? (forbes.com)</u>, Article "Will Barnes & Noble's Next Chapter Be Its Last?".

126. See web site: Forbes, <u>Will Barnes & Noble's Next Chapter Be Its Last? (forbes.com)</u>, Article "Will Barnes & Noble's Next Chapter Be Its Last?".

127. See web site: Forbes, <u>Will Barnes & Noble's Next Chapter Be Its Last? (forbes.com)</u>, Article "Will Barnes & Noble's Next Chapter Be Its Last?".

128. See web site: Identity Management, <u>Identity Management Lessons from the Barnes and Noble Breach (solutionsreview.com)</u>, Article "Identity Management Lessons from the Barnes and Noble Breach" by Ben Canner.

129. See web site: Identity Management, <u>Identity Management Lessons from the Barnes and Noble Breach (solutionsreview.com)</u>, Article "Identity Management Lessons from the Barnes and Noble Breach" by Ben Canner.

130. See web site: Identity Management, <u>Identity Management Lessons from the Barnes and Noble Breach (solutionsreview.com)</u>, Article "Identity Management Lessons from the Barnes and Noble Breach" by Ben Canner.

131. See web site: c|net Tech, <u>Barnes & Noble to add NFC chips to Nooks (cnet.com)</u>, Article "Barnes & Noble to add NFC chips to Nooks".

132. See web site: c|net Tech, <u>Barnes & Noble to add NFC chips to Nooks (cnet.com)</u>, Article "Barnes & Noble to add NFC chips to Nooks".

133. See web site: c|net Tech, <u>Barnes & Noble to add NFC chips to Nooks (cnet.com)</u>, Article "Barnes & Noble to add NFC chips to Nooks".

134. See web site: folio3, <u>Workflow Automation Services & Solutions | Folio3</u>, Article "BARNES & NOBLE".

135. See web site: folio3, <u>Workflow Automation Services & Solutions | Folio3</u>, Article "BARNES & NOBLE".

FIVE

Jennifer Doudna's Trend
In Genetic Engineering

There is a paradigm shift ...

*From a practice of gene therapy that treats genetic
diseases in specific individuals,*

*To a practice of germline enhancement that changes
heritable characteristics in future generations.*

This chapter is about a paradigm shift that biochemist Jennifer Doudna discerns in genetic engineering. The shift moves from gene therapy to germline enhancement, that is, from treating disease in individuals to preventing disease in multiple generations. In genetic engineering, I see treatment experiencing a bifurcation into healing humans and designing humans. In this chapter, I also compare the paradigm shift discerned by Goudna in genetic engineering with the paradigm shifts discerned by Gartner and Wolfgang Giegerich in technology and psychology respectively.

The main sources of information that shape my thinking about genetic engineering are:

- "A Crack In Creation: Gene Editing and the Unthinkable Power to Control Evolution" co-authored by Jennifer A. Doudna and Samuel H. Sternberg[1]
- "Editing Humanity: the CRISPR Revolution and the New Era of Genome Editing" by Kevin Davies[2]
- "mRNA Technology Gave Us the First Covid-19 Vaccines. It Could Also Upend the Drug Industry" by Walter Isaacson[3]
- "The Code Breaker: Jennifer Doudna, Gene Editing, and the Future of the Human Race" by Walter Isaacson[4]

Although "A Crack In Creation" has two authors, it is written in the first person singular number, and it is written primarily about Jennifer Doudna's experiences. So, I cite Doudna in references to the book. In "A Crack In Creation" Doudna describes a paradigm shift in genetic engineering.

Here is how I summarize the shift:

There is a paradigm shift ...

*From a practice of gene therapy that treats genetic
diseases in specific individuals,*

*To a practice of germline enhancement that changes
heritable characteristics in future generations.*

Highlights of DNA History

Before describing Doudna's paradigm shift, it may be helpful to give a brief background about highlights in the history of DNA. DNA is an acronym for Deoxyribose Nucleic Acid, which is the name of the molecule that contains genetic instructions for living organisms.[5] One highlight in the history of DNA is the discovery of the molecular structure of DNA. In 1953, scientists James Watson and Francis Crick discovered that the molecular structure of DNA is shaped like a double helix, that is, a twisted ladder.[6] That discovery gave rise to modern molecular biology, which is largely concerned with understanding how genes control the chemical processes within cells.[7] The double helix contains four types of bases, known as adenine (A), cytosine (C), guanine (G), and thymine (T).[8] The two strands of the double helix are held in place by bonds that connect the bases. Adenine bonds with Thymine, and Cytosine bonds with Guanine.[9] Each pair fits at right angles to the sides of the twisted ladder. Given the sequence of the bases in one strand, the bases of the other strand are automatically determined, which means that when the two chains are separated, each serves as a template for a complementary new chain.[10] The sequence of the bases on the strands of the ladder provide instructions for assembling protein and Ribonucleic Acid (RNA) molecules.

Another highlight in DNA history was the Human Genome Project (HGP). The HGP was about an international, research program whose goal was the complete mapping of all the genes in the human body. All human genes taken together are called the human "genome".[11] The scientists sequenced and mapped the genome of Homo Sapiens. The HGP took 13 years, from 1990 to 2003, to develop the ability to read nature's complete genetic blueprint for building a human being.[12]

HGP researchers deciphered the human genome in three major ways:[13]

1. Determine the sequence of the bases in the human genome's DNA
2. Make maps that show locations of genes for major sections of chromosomes, and
3. Produce linkage maps, so that inherited traits for genetic disease can be tracked over generations.

The HGP sequenced three billion DNA letters in the human genome.[14]

Gene Therapy in the 20th Century

In "*A Crack In Creation*" Jennifer Doudna explains how the Human Genome Project facilitated gene therapy during research in the 20th century:[15]

> "Since the completion of the Human Genome Project, the process of DNA and whole-genome sequencing has become staggeringly quick, cheap and effective. Scientists have precisely identified well over four thousand different kinds of DNA mutations that can cause genetic disease. DNA sequencing can tell individuals if they're at elevated risk of developing certain cancers, and it can help tailor specific treatments to best match the genetic backgrounds of different patients.
>
> …..
>
> Yet, while genome sequencing represents a huge development in the study of genetic disease, it is ultimately a diagnostic tool, not a form of treatment. It has allowed us to see how genetic diseases are written in the language of DNA, but it leaves us

powerless to change that language."

Since 20th century treatments for genetic diseases, such as sickle cell disease, involved laborious methods, scientists sought to use their knowledge of DNA to repair the defective genes directly.[16] Gene therapy is about repairing the gene itself to permanently reverse the disease in one person. One issue was locating the defective gene, another was the issue of how to administer the repair. The HGP facilitated location of defective genes. Scientists needed to figure out how to administer repairs. Scientists realized that viruses are effective in transferring genetic information. Like a Trojan horse, a virus is good at splicing their own genetic information into cells of any type.[17] Viruses know, not merely how to get into cells, they also know how to make their genetic code take up residence in cells.[18] So scientists set out to use viruses to deliver genetic material into the genome of infected cells. In the late 20th century, scientists used this approach as a form of gene therapy to cut and paste segments of DNA.[19] The use of viruses as a vehicle for conducting gene therapy was successful in repairing genes in heritable diseases such as cystic fibrosis and hemophilia. However, scientists found that gene therapy could be as harmful as it could be useful. Sometimes, there were mistakes in targeting defective genes.[20] They needed a way to repair defective genes without the risk of splicing genetic material into the wrong target. They looked into the possibility of using RNA to guide the editing of defective genes.[21]

Following the completion of the Human Genome Project, gene therapy experienced a flurry of activity. There were experiments, trials, successes, failures and lawsuits. Overall, the excitement about the potential of gene therapy outweighed its reliability. There were issues of safety because targeting genes was not a precise procedure and targeting mistakes put the safety of patients in jeopardy. This is how Kevin Davies described the status of gene therapy in his book "*Editing Humanity*":[22]

"(D)espite hundreds of millions of dollars lavished on hundreds of gene therapy trials involving thousands of patients and volunteers, 'new hopes cynically turned to ashes, dramatic claims to sad farce. Gene therapy itself was on life support.

…..

With some sober reflection, many of the setbacks could be understood. After all, viruses did not evolve simply to be used at our beck and call as delivery drones. As one gene therapy expert said: 'We underestimated the fact that it took billions of years for the viruses to learn to live in us —- and we were hoping to do it in a five-year grant cycle.' There was also the complication of our immune system, which is designed to combat foreign agents such as viruses. The human body isn't going to automatically give billions of recombinant viruses a pass just because they mean well.

It has been a long haul back to respectability and success for gene augmentation therapy. The roller-coaster ride follows the … inflated expectations of the 1990s, the trough —— or abyss —— of disillusionment at the turn of the century, followed by the slope of enlightenment. What's been holding up the field is not a lack of suitable targets —- we have an encyclopedic catalogue of thousands of eligible Mendelian genetic diseases —- but the capability of delivering the therapeutic gene safety and effectivity. Researchers had to go back to the drawing board, focusing on viral delivery and safety."

Germline Enhancement in the 21st Century

Walter Isaacson was writing a biography of Jennifer Doudna when Doudna and her colleague Emmanuelle Charpentier were awarded the 2020 Nobel Prize for Chemistry. Isaacson wrote an article in TIME Magazine to describe his research on Doudna's work. Here is how he described the superiority of mRNA (RNA guide) gene editing over the DNA gene editing potential of CRISPR, the gene editing tool that was being considered for use in producing a Coronavirus vaccine at the time:[23]

"Doudna and a fellow scientist had invented an RNA-guided, gene-editing tool called CRISPR, for which they won the Nobel Prize in Chemistry in 2020.

…..

An mRNA vaccine has certain advantages over a DNA vaccine, which has to use a re-engineered virus or other delivery mechanism to make it through the membrane that protects the nucleus of a cell. The RNA does not need to get into the nucleus. It simply needs to be delivered into the more-accessible outer region of cells, the cytoplasm, which is where proteins are constructed."

CRISPR (**C**lustered **R**egularly **I**nterspaced **S**hort **P**alindromic **R**epeats) is a gene editing tool whose purpose is to alter the sequence of DNA to accomplish a goal such as correcting a mutation that can result in a genetic disease.

This is how Jennifer Doudna explains CRISPR:[24]

"CRISPR can be described as a pair of designer molecular scissors because of its core function: to home in on specific twenty-letter DNA sequences and cut apart both strands of the double helix. Yet the types of gene-editing outcomes that scientists can achieve with this technology are remarkably diverse. For this reason, it might be better to describe CRISPR not as scissors but as a Swiss army knife, a tool with a panoply of functionalities that all stem from the action of a singular molecular machine.

The simplest use of CRISPR is also the one that's most widely employed: have it cut a specific gene and then allow the cell to repair the damage by reconnecting the strands. ... Even though scientists can't control the exact way that the DNA is repaired in this particular use of CRISPR, they've realized how useful this type of gene editing can be.

Genes are, after all, just carriers of information, like the blueprints of a house: the goal of gene editing is not merely to alter the blueprints, but to change the form of the structure that gets built."

Doudna further explains:[25]

"As long as the genetic code for a particular trait is known, scientists can use CRISPR to insert, edit, or delete the associated gene in virtually any living plant's or animal's genome. This process is far simpler and more effective than any other gene-manipulation technology in existence. Practically overnight, we have found ourselves on the cusp of a new age in genetic engineering and biological mastery – a revolutionary era in which the possibilities are limited only by our collective imagination."

In the 21st century, when CRISPR became available, scientists were able to use it for germline enhancement. The intention of germline enhancement is to prevent a genetic disease by editing the mutation before it manifests as a disease in a person's body. Here are examples of what scientists have accomplished with germline enhancement:[26]

- Scientists have applied CRISPR to lung cells to correct cystic fibrosis, and to blood cells to correct sickle cell disease.
- Scientists have also used CRISPR to edit and repair mutations in stem cells which can then be influenced to transform into virtually any cell or tissue type in the body.
- In addition to simply splicing apart DNA, and inserting new sequence into the target genome, scientists can now also deactivate the gene, rearrange sequences of genetic code, and even correct single-letter mistakes.

Between the 20th century and the 21st century, the intention behind genetic engineering for humans had changed. The intention behind 20th century gene therapy was to provide treatment for genetic disease after the disease manifested in humans, and the manifestation could be well into adulthood. The intention behind 21st century germline enhancement is to prevent genetic disease by altering the DNA of embryos. It turns out that germline enhancement has greater potential than just disease prevention. It also has the potential for producing designer babies, that is, babies who have no genetic disease, but whose doctors and parents choose germline enhancement for improving human features like physical appearance, intelligence and athletic capability.

As the technology that supports genetic engineering changes, experts adjust their expectations about what they can accomplish with the technology. As their use of technology becomes more sophisticated, I see a bifurcation of treatment into healing and designing. The 20th century treatment was gene therapy intended to heal genetic diseases already manifesting in humans. The 21st century treatment is germline enhancement intended to prevent disease, by editing genes in embryos. The realization that germline enhancement can also design babies, by modifying genes for physical appearance and certain capabilities, puts Homo Sapiens on a path to generating a new version of humanity.

Germline enhancement raises questions about the ethical uses of gene editing tools. In addition to recommending ethical caution, Doudna has arranged for scientists and others to debate ethical uses of germline enhancement, because she is of the view that humanity needs to establish ethical guidelines for use of technologies like CRISPR. In his book "*Editing Humanity*" Kevin Davies reports that Jennifer Doudna, Emmanuelle Charpentier, other scientists and the public participated in a conference about ethics. Davies informs us that in 2015, there was a conference about the ethics of genome editing at the National Academy of Sciences.

"For three days, scientists, physicians, ethicists, and philosophers debated the danger and potential application of germline editing. Although in the minority, some spoke out strongly in favor of germline editing, at least in principle. Manchester University philosopher John Harris quipped, 'If sex had been invented, it would never have been permitted or licensed ... it's far too dangerous!' But the most electrifying moment came not on stage but from the audience. Sarah Gray rose up and spoke tearfully about the birth of her son with anencephaly, who suffered seizures for a week before his death. 'If you have the skills and the knowledge to eliminate these diseases, then freakin' do it!' she pleaded." [27]

Comparing Doudna's Trend In Genetic Engineering With Gartner's Trend In Technology

Jennifer Doudna's trend in genetic engineering parallels Gartner's trend in technology. I describe Gartner's trend in detail in Chapter 1: "Gartner's Trend In Technology".

As a reminder, this is how I summarize Gartner's trend in technology:

There is a paradigm shift ...

From a model of Technology-Literate People,
where individuals acquire knowledge about digitalization,

To a model of People-Literate Technology,
where algorithms acquire knowledge about people.

In genetic engineering of the 20th century, the situation was primarily one of human scientists learning about the available technology and creating additional technology for addressing genetic disease. In the 21st century, genetic engineering has spawned a subfield in which there is a growing number of Machine Learning algorithms that are learning about people. What follows are three examples of algorithms that are learning about people in genetic engineering:

- **SPROUT** (CRISPR Repair Outcome) is an algorithm that is people-literate in the sense that it is learning the behavior of people, that is, the mistakes that scientists are making during gene editing. Using Machine Learning, scientists have created an algorithm which predicts the type of mistakes that are likely during CRISPR editing.[28] When CRISPR is used in an edit, there is a strand of molecules, a guide RNA which guides the DNA-slicing protein Cas9 to the target DNA. On arrival at the target DNA, the guide RNA binds to the DNA and Cas9 makes a cut so that the new DNA can be inserted, or the old DNA deleted. CRISPR based edits involve risk. Some edits may have unexpected results.[29] The Machine Learning algorithm named CRISPR Repair Outcome (SPROUT) uses large volumes of data about such mistakes, related to human cells, to learn sequence patterns prone to erroneous DNA editing. SPROUT is able to predict the likelihood of an error, and whether the error would damage the gene. The predictive power of the algorithm is helpful to scientists. When they have options about where to edit a mutation in DNA gene, they can choose the option with the least probability of error and damage.[30]

- **inDelphi** is an algorithm that is acquiring people-literacy in the area of people's biology, by learning how to predict the repair of human cells after gene editing is complete.[31] inDelphi is a Machine Learning algorithm whose goal is help scientists using CRISPR to predict the outcome of gene editing. This is also an algorithm that is people-literate technology because it is learning about the propensity for human biology to recover after gene editing. As a computational model, inDelphi predicts the mixture of outcomes from small insertions and deletions in human genomes resulting from CRISPR-induced edits of genes. inDelphi synthesizes known biological mechanisms of DNA repair with Machine Learning to make predictions. The scientists believe that inDelphi can enable an improvement in the efficiency of DNA repair pathway, whose outcomes can be predicted and thereby controlled.

- **FORECasT** (**F**avored **O**utcomes of **R**epair **E**vents at **Cas9 T**argets) is an algorithm that is becoming people-literate about the repair of the human cells after gene editing is complete. Scientists created a Machine Learning algorithm named FORECasT that predicts the outcomes of gene editing repairs.[32] Using a library of numerous guide-RNAs and a dataset containing records of more than one billion repairs in various cell types, they showed that the majority of repairs are either single base insertions, small deletions, or longer deletions called microhomology-mediated deletions. They also showed that completion of repair is based on specific sequences that exist at the genome site where the gene is edited. The algorithm was able to use the sequences that determine each repair to predict editing outcomes. Accurate predictions of sequence repair allow researchers to computationally predict the precise guide RNAs that will produce the best healing results for human patient mutations.

SPROUT, inDelphi and FORECasT are all Machine Learning algorithms that are being prepared for commercialization. They mine data collected from gene editing activity to learn about the behavior of people and the bodies of people. Then, the algorithms use their people-literacy to make predictions about the outcomes of gene editing. These gene editing algorithms provide evidence of Gartner's trend toward people-literate technology because they are learning about people. SPROUT is becoming literate about the behavior of scientists who conduct gene editing, while inDelphi and FORECasT are becoming literate about the people's bodies as they heal after gene editing.

Doudna's trend in genetic engineering is moving from gene therapy, where scientists are literate about gene editing technology, to germline enhancement, where algorithms are becoming literate about patients and scientists. Technology enables the collection of volumes of data large enough for algorithms to become literate about people. Gartner's trend in technology moves from technology-literate people to people-literate technology. In a similar trend, SPROUT, inDelphi and FORECasT all indicate that algorithms are becoming literate about people.

The value that Machine Learning adds to gene editing is that as algorithms mine greater volumes of data, they learn to be more precise in detecting patterns. In Gartner's sense of the expression, SPROUT is a people-literate technology. SPROUT is useful to gene editing because it can acquire knowledge about mistakes that people make in gene editing. inDelphi and FORECasT are also people-literate technology in the sense that they are learning how human bodies become repaired after being gene edited.

Comparing Doudna's Trend In Genetic Engineering With Giegerich's Trend In Technology

Jennifer Doudna's trend in genetic engineering also parallels Wolfgang Giegerich's trend in psychology. I describe Giegerich's trend in detail in Chapter 2: "Wolfgang Giegerich's Trend In Psychology".

Here is a reminder of how I summarize Giegerich's trend in psychology:

There is a paradigm shift …

From a focus on the semantical level of psychology,
where individuals engage in the individuation process,
a goal-seeking effort to differentiate their minds from the
unconsciousness of their communities,

To a focus on the syntactical level of psychology,
where human culture engages in the interiorization process,
an intellectual discipline of interpreting phenomena
that emerge in the world.

Giegerich's trend is from the semantical level to the syntactical level of psychology. When scientists performed gene therapy in the 20th century, they were aware of the connection between certain genes and specific diseases. When a disease manifested in humans, there was research about the cause of the disease. If the cause was determined to be genetic, it was classified as a disease associated with a particular gene. Their approach was one of finding a treatment or cure for particular diseases. Genetic diseases were researched and treated individually. In the minds of scientists at the time, genetic diseases were separate, and therefore handled individually. Gene therapy was about treating genetic diseases individually after the diseases manifested in human bodies. That approach puts consciousness of genetic engineering on Giegerich's semantical level of psychology. Scientists and the public acquired knowledge of genetic disease on a gene-by-gene basis, adding knowledge incrementally. Their consciousness was oriented toward accumulating knowledge of genes incrementally, block by block style.

In a 13-year period that crosses over from the 20th into the 21st century, the Human Genome Project (HGP) produced a map of the human genome. The HGP decoded and mapped knowledge of all the genes that make up the human genome. The blueprint for the genetic makeup of Homo Sapiens became available in the public domain. The web site for National Human Genome Research Institute makes the following statement: [33]

> "The human genome contains approximately 3 billion of these base pairs, which reside in the 23 pairs of chromosomes within the nucleus of all our cells. Each chromosome contains hundreds to thousands of genes, which carry the instructions for making proteins. Each of the estimated 30,000 genes in the human genome makes an average of three proteins."

The mapping of the sequence of DNA letters in the human genome gave scientists a new cohesive and holistic outlook on genetic engineering. The blueprint was not just additional knowledge about more genes; it was a comprehensive, structured mapping that gave genetic knowledge a new arrangement. Scientists used knowledge of the DNA mapping as a basis for taking a new approach. Instead of treating diseases after they manifest in humans, scientists adopted a futuristic approach. The HGP was instrumental in a transformation of the way scientists view genes. Before the HGP, scientists had an approach of recovery from genetic disease. After the HGP, scientists adopted an approach of prevention of genetic disease. The shift from recovery to prevention marked a new outlook in genetic engineering.

With help from the HGP, scientists began to take a new approach to genetic engineering. The genetic knowledge that had been available incrementally, building block style in the 20th century, had come together and been enriched to become a full-blown genetic blueprint for building Homo Sapiens. The sequence of DNA in the human genome created a new arrangement of knowledge about human molecular biology. Scientists adopted a new approach to correcting the mutations that cause genetic diseases. Gene editing technology was applicable across all the genes in the human genome. Consciousness of gene therapy, which involves treatment for individual gene mutations, morphed into consciousness of gene editing, which involves applying technology to any gene in the human genome. Consciousness had been restructured from an outlook of collecting knowledge about genes, one at a time, to an outlook of using technology to edit any mutation in the human genome. As scientists shared their knowledge of genetic disease with the public, the restructure of consciousness spread to the public domain.

Still later, in the 21st century, scientific research revealed that by editing stem cells in an embryo, they create a germline enhancement. Scientists not only prevent a genetic disease in one person, but also prevent the disease in future descendants of that person. Germline enhancement led to a further restructuring of consciousness in scientists, as well as the public. What makes this restructuring monumental is that germline enhancement enables humanity, not just to prevent genetic diseases, but also to select preferred characteristics in babies. The technology exists for scientists and parents to select preferred characteristics in embryos, for example, physical appearances, with the possibilities of also selecting athletic ability and intelligence. Scientists have the knowledge and the skills to build a new version of Homo Sapiens.

The restructured knowledge of genes coupled with the new capability of changing the genes of future generations put consciousness of genetic engineering on Giegerich's syntactical level of psychology. Doudna's trend in genetic engineering is consistent with Giegerich's trend in psychology. They both address changes in consciousness. Doudna's trend in genetic engineering plays out in the public arena, where there are publications of scientific papers, scientists being interviewed on television, postings in social media, as well as treatment and prevention of genetic disease in people. The publicity about changes in genetic engineering provides evidence of human consciousness being restructured during the trend. Doudna's trend gives visibility into Giegerich's subtler trend, which is not as well known in the public domain.

Regarding germline enhancement, Doudna states:

> "CRISPR gives us the power to radically and irreversibly alter the biosphere that we inhabit by providing a way to rewrite the very molecules of life in any way we wish." [34]

That outlook is definitely a restructuring of consciousness about genetic engineering. Scientists used to simply treat genetic diseases; now they are modifying future generations. What remains is for the international community to set standards about what genetic modifications are acceptable, and in what situations they are applicable. I believe the international community has a responsibility to set a code of ethics because of the enormity of the potential for germline enhancement to influence the future as Homo Sapiens sets out on a path for generating a new version of humanity.

Summary

Between the 20th century and the 21st century, there were three events that changed genetic engineering significantly:

1. Francis Crick and James Watson discovered the double-helix structure of DNA and how it influences chemical processes within cells.
2. A consortium of international scientists completed the Human Genome Project, which produced the genetic coding and mapping of the human genome.
3. Jennifer Doudna and Emmanuelle Charpentier won the 2020 Nobel Prize in Chemistry, for their development of the technology named CRISPR/Cas9 which can edit the human genome.

With the knowledge gathered from these three events, scientists can modify the human genome in ways that change the genes of future generations. The trends I compare in this chapter are Doudna's trend in genetic engineering, Gartner's trend in technology and Giegerich's trend in psychology. Together, these trends contribute to an overarching paradigm shift that puts Homo Sapiens on the cusp of creating a new version of humanity.

The mindset of scientists in the 20th century was to regard treatment in terms of healing humans, one genetic disease at a time. That mindset put consciousness of genetic engineering on Giegerich's semantical level of psychology. Scientists and the public acquired knowledge of genetic disease on a gene-by-gene basis, adding knowledge incrementally. Their consciousness was oriented toward accumulating knowledge of genes separately. At the turn of the century, the Human Genome Project created a new arrangement of knowledge about human molecular biology. In the 21st century, the mindset of scientists changed to regard treatment in terms of designing humans. Scientists acquired the knowledge and the skills to build a new version of Homo Sapiens by selecting physical appearance, and possibly intelligence. The restructured knowledge of genes, coupled with the new capability of changing the genes of future generations, put consciousness of genetic engineering on Giegerich's syntactical level of psychology.

The Human Genome Project provided knowledge of the entire human genome. Genetic engineering has the capability to edit any gene, not just for preventing disease, but also for influencing future generations. In addition, genetic engineering offers the potential for optionally enhancing human appearance, and gene-influenced capabilities. Homo Sapiens has within reach all that is needed to change human biology. What Doudna's work contributes to the overarching paradigm shift is the sophistication of the genetic engineering tool CRISPR/Cas9 to modify the genetic composition, and therefore the biology of humans. What Gartner's trend contributes is the Machine Learning technology that enables algorithms, such as SPROUT, inDelphi and FORECasT, to acquire knowledge about people, that is, scientists and patients, as they participate in genetic engineering. What Giegerich's trend contributes to the overarching paradigm shift is that humanity is shifting its outlook from eliminating disease to designing babies. That shift involves a restructure of consciousness.

Together, Doudna's trend in genetic engineering, Gartner's trend in technology and Giegerich's trend in psychology, all point to the fact that scientists have the knowledge and the skills to build a new version of Homo Sapiens.

NOTES:

1. See "*A Crack In Creation: Gene Editing and the Unthinkable Power to Control Evolution*" co-authored by Jennifer A. Doudna and Samuel H. Sternberg, First Mariner Books, 2018.
2. See "*Editing Humanity: the CRISPR Revolution and the New Era of Genome Editing*" by Kevin Davies, Penguin Books, 2020.
3. See "*mRNA Technology Gave Us the First Covid-19 Vaccines. It Could Also Upend the Drug Industry*" by Walter Isaacson, TIME Magazine, January 11, 2021.

4. See *"The Code Breaker: Jennifer Doudna, Gene Editing, and the Future of the Human Race"* by Walter Isaacson, Simon & Schuster, 2021.

5. See web site: Web site: NIH: Human Genome Institute, Deoxyribonucleic Acid (DNA) (genome.gov), Article "Deoxyribonucleic Acid (DNA)".

6. See web site: NIH: U.S. National Library of Medicine, The Discovery of the Double Helix, 1951-1953 | Francis Crick - Profiles in Science (nih.gov), Article "The Discovery of the Double Helix, 1951 – 1853".

7. See web site: NIH: U.S. National Library of Medicine, The Discovery of the Double Helix, 1951-1953 | Francis Crick - Profiles in Science (nih.gov), Article "The Discovery of the Double Helix, 1951 – 1853".

8. See web site: NIH: Human Genome Institute, Deoxyribonucleic Acid (DNA) (genome.gov), Article "Deoxyribonucleic Acid (DNA)".

9. See web site: NIH: Human Genome Institute, Deoxyribonucleic Acid (DNA) (genome.gov), Article "Deoxyribonucleic Acid (DNA)".

10. See web site: NIH: Human Genome Institute, Deoxyribonucleic Acid (DNA) (genome.gov), Article "Deoxyribonucleic Acid (DNA)".

11. See web site: NIH: National Human Genome Research Institute, The Human Genome Project, Article "The Human Genome Project".

12. See web site: NIH: National Human Genome Research Institute, The Human Genome Project, Article "The Human Genome Project".

13. See web site: NIH: National Human Genome Research Institute, The Human Genome Project, Article "The Human Genome Project".

14. See web site: NIH: National Human Genome Research Institute, The Human Genome Project, Article "The Human Genome Project".

15. See "*A Crack In Creation: Gene Editing and the Unthinkable Power to Control Evolution*" co-authored by Jennifer A. Doudna and Samuel H. Sternberg, p 15.
16. See "*A Crack In Creation: Gene Editing and the Unthinkable Power to Control Evolution*" co-authored by Jennifer A. Doudna and Samuel H. Sternberg, p 16.
17. See "*A Crack In Creation: Gene Editing and the Unthinkable Power to Control Evolution*" co-authored by Jennifer A. Doudna and Samuel H. Sternberg, pp 16 – 18.
18. See "*A Crack In Creation: Gene Editing and the Unthinkable Power to Control Evolution*" co-authored by Jennifer A. Doudna and Samuel H. Sternberg, p 18.
19. See "*A Crack In Creation: Gene Editing and the Unthinkable Power to Control Evolution*" co-authored by Jennifer A. Doudna and Samuel H. Sternberg, p 19.
20. See "*A Crack In Creation: Gene Editing and the Unthinkable Power to Control Evolution*" co-authored by Jennifer A. Doudna and Samuel H. Sternberg, p 20.
21. See "*A Crack In Creation: Gene Editing and the Unthinkable Power to Control Evolution*" co-authored by Jennifer A. Doudna and Samuel H. Sternberg, p 21.
22. See "*Editing Humanity: the CRISPR Revolution and the New Era of Genome Editing*" by Kevin Davies, pp 144 – 145.
23. See "*mRNA Technology Gave Us the First Covid-19 Vaccines. It Could Also Upend the Drug Industry*" by Walter Isaacson, TIME Magazine, January 11, 2021.
24. See "*A Crack In Creation: Gene Editing and the Unthinkable Power to Control Evolution*" co-authored by Jennifer A. Doudna and Samuel H. Sternberg, pp 101 – 102.
25. See "*A Crack In Creation: Gene Editing and the Unthinkable Power to Control Evolution*" co-authored by Jennifer A. Doudna and Samuel H. Sternberg, p xiii.
26. See "*A Crack In Creation: Gene Editing and the Unthinkable Power to Control Evolution*" co-authored by Jennifer A. Doudna and Samuel H. Sternberg, p 100.

27. See *"Editing Humanity: the CRISPR Revolution and the New Era of Genome Editing"* by Kevin Davies, pp 254 – 255.

28. See web site: Stanford Medicine, <u>CRISPR algorithm predicts how well gene editing will work - Scope (stanford.edu)</u>, Article "CRISPR algorithm predicts how well gene editing will work" by Hanae Armitage

29. See web site: Stanford Medicine, <u>CRISPR algorithm predicts how well gene editing will work - Scope (stanford.edu)</u>, Article "CRISPR algorithm predicts how well gene editing will work" by Hanae Armitage.

30. See web site: Stanford Medicine, <u>CRISPR algorithm predicts how well gene editing will work - Scope (stanford.edu)</u>, Article "CRISPR algorithm predicts how well gene editing will work" by Hanae Armitage.

31. See web site: TheScientist, <u>Could AI Make Gene Editing More Accurate | The Scientist Magazine® (the-scientist.com)</u>, Article "Could AI Make Gene Editing More Accurate" by Ashley Yeager. **The papers being referenced on the web site:** M.W. Shen et al., "Predictable and precise template-free CRISPR editing of pathogenic variants," *Nature*, 563:646–51, 2018. F. Allen et al., "Predicting the mutations generated by repair of Cas9-induced double-strand breaks," *Nat Biotechnol*, 37:64–72, 2019.

32. See web site: TheScientist, <u>Could AI Make Gene Editing More Accurate | The Scientist Magazine® (the-scientist.com)</u>, Article "Could AI Make Gene Editing More Accurate" by Ashley Yeager. **The papers being referenced on the web site:** M.W. Shen et al., "Predictable and precise template-free CRISPR editing of pathogenic variants," *Nature*, 563:646–51, 2018. F. Allen et al., "Predicting the mutations generated by repair of Cas9-induced double-strand breaks," *Nat Biotechnol*, 37:64–72, 2019.

33. See web site: National Human Genome Research Institute, <u>Human Genome Project FAQ</u>, Article "What is a genome?".

34. See *"A Crack In Creation: Gene Editing and the Unthinkable Power to Control Evolution"* co-authored by Jennifer A. Doudna and Samuel H. Sternberg, p 119.

Project Management Institute's Trend In Project Management

There is a paradigm shift ...

From project management in traditional enterprises
that have a lesser focus on building a digital culture,

To project management in gymnastic enterprises
that have a greater focus on building a digital culture.

The Project Management Institute (PMI) is an organization that publishes standards for managing projects. The Institute also conducts training for aspiring project managers, who want to acquire professional credentials. At the time of writing in 2021, the PMI has over 600,00 members and over 300 chapters worldwide.[1] In this chapter, I describe the PMI's trend in project management, from traditional enterprises to gymnastic enterprises. I explain my proposal that project management is bifurcating into data management and strategy management. In addition, I compare the PMI's trend with Gartner's trend in technology, as well as Giegerich's trend in psychology.

The main sources of information for this chapter are:

- PMI's *Pulse of the Profession* 2021.[2]
- "A Guide to the Project Management Body of Knowledge (PMBOK® Guide)" Seventh Edition.[3]
- Web Site: Project Management Institute.[4]
- Web Site: International Project Management Association.[5]

Trend in Project Management

According to PMI's 2021 *Pulse of the Profession*, gymnastic enterprises are emerging from traditional enterprises. The Coronavirus pandemic forced many organizations to allow employees to work remotely, and business needs are changing as enterprises compete in an increasingly digital culture. In support of the digital culture, Artificial Intelligence is performing a growing percent of project and portfolio management tasks.[6]

In the 20th century, project management software accommodated the Waterfall Methodology, and used tools like Gantt Chart, Work Breakdown Structure Diagram, Milestone roll-up, Critical Path Analysis, and Project Review Evaluation Technique (PERT).[7] The crux of project management used to be balancing work, cost and time for a successful project outcome. Project management tools were available for setting up milestones, assigning tasks, establishing timelines, and monitoring progress. To use these tools effectively, project managers had to become literate about the technology upon which the tools were built.

In the 21st century, project management software now accommodates Scaled Agile Framework (SAFe)[8] and uses tools like Zoho Sprints, airfocus and ProjectWise.[9] The essence of project management in SAFe is carving up work into portions that can be implemented quickly in iterations of Viable Work Products. As project management evolves in the digital culture, tools used in both the Waterfall Methodology and the SAFe include Artificial Intelligence based algorithms that are capable of becoming literate about people, such as software developers and project managers. The PMI's trend in project management moves from traditional enterprises to gymnastic enterprises, which readily embrace digital culture.

PMI's Pulse of the Profession

In "Pulse of the Profession" 2021, the PMI identified a project management trend that I summarize as follows:

There is a paradigm shift …

*From project management in traditional enterprises
that have a lesser focus on building a digital culture,*

*To project management in gymnastic enterprises
that have a greater focus on building a digital culture.*

The PMI's "Pulse of the Profession" for 2021 was composed from an online survey of 3,950 project professionals who were interviewed in October and November 2020.[10] Their projects represented a variety of industries in geographic areas that include North America, Europe, Asia Pacific, Sub-Saharan Africa, Latin America, South Asia, Middle East/ North Africa, and China.[11]

The PMI discerned a new type of enterprise which it labels "gymnastic enterprise" emerging from traditional enterprises. The survey revealed that professionals regard digital transformation as the factor that had the most significant impact on their enterprises in the past year. Here is an excerpt from the survey results:[12]

> "PMI's 2021 *Pulse of the Profession®* survey reveals the emergence of what we call the gymnastic enterprise. These organizations and their project teams combine structure, form, and governance with the ability to flex and pivot—wherever and whenever needed.
>
> Gymnastic enterprises empower their people to master new ways of working by applying the right approach in the right way at the right time to deliver what's needed. Sometimes that's agile; other times it's hybrid or waterfall. And they help their people work more effectively by using a variety of powerful tech-enhanced approaches like complex problem-solving techniques and on-demand, microlearning apps. It's all about figuring out what works for the project.
>
> ...
>
> By enabling their people to become changemakers, gymnastic enterprises are better able to sense and respond to shocks, drive change, and face the future, knowing that they have the mindset, skills, and tools that it takes to win.
>
> ...
>
> The world has no doubt seen extraordinary change in the last few years. But it turns out that was a mere ripple compared to

the tsunami of disruption we saw in 2020. As the (COVID-19) pandemic upended the world, it accelerated new ways of working and delivering value that had been gathering steam for some time. It accelerated the pace and scale of digitalization exponentially, with a big impact on talent and the need to upskill and reskill."

Based on the 2021 *Pulse of the Profession®* survey, the PMI identifies two types of enterprise: gymnastic enterprise and traditional enterprise:[13] This is how the PMI distinguishes between the two types of enterprise. Gymnastic enterprises are more likely than traditional enterprises to:

- Focus on building a digital culture
- Use different types of technology
- Embrace digital transformation
- Use cloud solutions, the Internet of Things (IoT), Artificial Intelligence (AI), and 5G mobile Internet
- Use tech-enabled ways of working, such as, complex problem-solving techniques, AI-driven tools, microlearning apps, and career assessment tools, and
- Prioritize a culture that's receptive to change and which fosters innovation.

Traditionally, project management and strategy management have been separate roles, but the current trend is that project management is being expanded to include organizational strategy.[14] Project managers are expected to understand the relationships among program, portfolio, and project management, because the relationships allow organizations to see how individual projects are related to each other and fit within the overall strategic goals of the organization.[15]

A significant project management trend, noted by the PMI, is the decline in popularity of Project Management Offices (PMOs).[16] One explanation being given for the decline is that PMOs have been part of hierarchical organizational structures which are now being flattened.[17] That I find understandable. Another explanation being offered is that lack of executive involvement makes it hard for PMOs to understand strategic goals.[18] That second explanation does not stand up to scrutiny. I have to wonder if the solution fits the problem. Is it really the project managers' lack of involvement in strategy that makes it hard for PMOs to understand strategic goals, or is it that executives find it difficult to articulate their strategic goals?

A Project Management Survey conducted in 2019 by Klynveld Peat Marwick Goerdeler (KPMG), Australian Institute of Project Management (AIPM) and International Project Management Association (IPMA) indicated that only 30% of the organizations surveyed believe their capabilities in change management are effective, so they are looking to project managers to drive change, for example, digital transformation.[19] Here again I have to wonder if the solution fits the problem. Does the fact that change managers are failing to effectively drive change in digital transformation mean that project managers become responsible for taking over change management?

In my observation, credentialed project managers tend to be people with strong organizational skills. Those with formal project management credentials tend to have hard, technical skills as well as soft, people skills. I believe that many credentialed project managers are capable of supporting strategic goals and capable of taking over the role of change management. However, I do not believe they would be effective if they were to be subservient to executives who have difficulty articulating their strategic goals, nor would they be effective if they were subservient to change managers who do not grasp digital transformation. In my opinion, credentialed project managers are up to the tasks of supporting strategic goals and driving change management, but not if they are used to prop up executives with communication challenges, and digitally ignorant change managers.

I see a bifurcation in the role of project manager. In the 20th century, a project manager's role used to be organizing projects according to a chosen methodology, performing the sustained number-crunching in-volved in balancing work, cost and time, while providing project updates to stakeholders. In the 21st century, there are Artificial Intelligence based technologies that can perform the organization of projects according to a selected methodology, as well as the relevant number-crunching. Moreover, there are Machine Learning algorithms that are capable of learning about project management by mining large volumes of data col-lected about project management.

The role of project manager is now more about decision-making regard-ing organizational strategy embodied in portfolios. In my opinion, project management is experiencing a bifurcation into data management per-formed by algorithms, and strategy management performed by human project and portfolio managers. Here are examples of technologies that are performing what used to be project management responsibilities:

- Ubisoft's Commit Assistant
- PPM Express' PPM Insight, and
- IBM's Watson Studio.

Ubisoft – Commit Assistant

In traditional project management, people are responsible for finding errors (bugs) in computer code and correcting them. That activity is shifting from people to algorithms. Ubisoft is a French video game company that has its headquarters in Paris, and product development sites in many cities worldwide.[20] Ubisoft has a Machine Learning system named Commit Assistant which uses algorithms to identify errors before a project gets to the testing phase of project management. Commit Assistant learned error detection by examining ten years of computer code to identify errors that have been found and corrected in the past.[21] With this knowledge of errors made and corrected by developers, Ubisoft is able to reduce development and test-ing time during project management and prevent errors from being deployed into the final delivery of video games.[22]

Ubisoft's Commit Assistant includes algorithms that are becoming literate about people, in particular software developers, who write computer code, and testers who check the accuracy of the code before new technology-based products are determined fit for market. Based on historical data about errors, the algorithms are learning to detect new errors that developers make while coding, so that the errors can be corrected before launch of video games. The algorithms are learning how developers make errors and what corrections correspond to those errors.

Ubisoft fed Commit Assistant ten years of historical data about video game code from Ubisoft's Library.[23] Machine Learning capabilities enable Commit Assistant to learn the history of errors that have been identified, what was done to correct the errors and to predict future mistakes that developers might make while writing code. Commit Assistant categorizes the code and uses its algorithm to match newly written code against the ten-year history of errors. When there is a match, the Commit Assistant produces a notification of the error. As the volume of historic data grows, the Commit Assistant is expected to improve its performance in finding errors.[24] Thanks to Machine Learning capabilities, Commit Assistant can learn where mistakes have been made, what corrections were made and predict future mistakes developers might make.[25] When a coding error is identified, Commit Assistant can find the most likely cause and suggest fixes for programmers to implement. Over time, Ubisoft estimates that Commit Assistant will save 20 percent of development time.[26]

Commit Assistant does not detect errors the way that human developers and testers do; it restructures the whole process of error detection and correction. Humans detect errors incrementally, expanding their historical knowledge of error detection one at a time. Commit Assistant studies the history of errors, categorized them, and use its algorithm to compare new software with the history of errors. On finding a match, the algorithm searches the history for a corresponding correction for the error, then generates a notification about the error detected and the corresponding correction. Commit Assistant restructures the error detection process by leveraging the ten-year history of errors and corrections. With the continued accumulation of historic data, the algorithm is expected to improve its performance in error detection. In the Ubisoft context, the process of error detection has been restructured by a Machine Learning system. The way in which Ubisoft employees regard error detection has been restructured. It is no longer a search for logical flaws in lines of code. It is about finding a match between newly written software and the history of errors. The introduction of the algorithm did not merely speed up the process for detecting errors; the algorithm restructured the way humans think about error detection. The human thinking about error detection transitioned from humans finding errors one-by-one, to humans experiencing an algorithm-induced restructure of thinking about error detection.

PPM Express - PPM Insights

PPM Express is a Project Portfolio Management Software Company that has offices in Seattle, Washington in the United States, and a development site in Ukraine. PPM Express markets PPM Insights as Machine Learning technology for mitigating project risks and increasing Project Managers' productivity.[27] PPM Insights collects project data from projects which reside in multiple project management tools including **Microsoft Planner, Azure DevOps, Jira Software, and Microsoft Project Online.**[28] By assembling data from multiple projects into one portfolio on one platform, PPM Insights provides a single interface for managing and tracking of all the project issues and anomalies detected in the portfolio of projects.

PPM Insights integrates project data collected from multiple tools, and creates a portfolio of projects, so that an enterprise has visibility into the totality of multiple projects in one place.[29] Project data includes tasks, assignments, budget, resources, risks and issues pertaining to all the projects in the portfolio. PPM Insights enables both project managers and customers to engage in simulations, in which they change project variables to find out possible outcomes before committing to changes.

PPM Insights' Machine Learning capabilities in project and portfolio management enable forecasts of project statuses as well as continuous detection of anomalies, variances and problems in the portfolio.[30] With these capabilities, PPM Insights allows project managers and portfolio managers to address project issues and reduce project risks proactively.[31] PPM Insights structures portfolio data for use in Machine Learning models and identifies key parameters for detecting variances that pose risk to projects.[32] Project data are processed with Machine Learning algorithms to identify key parameters that are used to detect variances. PPM Insights analyzes the portfolio data while searching schedule, budget and resources for variances that pose risk.[33] PPM Insights uses Machine Learning algorithms to establish a customizable dashboard to provide visibility into multiple projects in one portfolio repository.[34] When variances are found, PPM Insights displays the risks on the dashboard, generates action plans for mitigating the risks and sends notifications to project managers. Project managers mitigate the risks by using modules in PPM Insights.

Resource Management Software is PPM Insights' module for guiding the mitigation of risks posed by work overloads and burn-outs. Project managers may respond to notifications about variances in resource utilization by allocating available resources when they are needed the most. By providing a dashboard, PPM Insights gives project managers visibility into the status of their projects. By generating an action plan for each project, PPM Insights enables project managers to increase their productivity by proactively monitoring and mitigating risks in their projects.[35] Project managers benefit from early detection of variances that pose risk to projects, so their project teams can take preventive action.[36] PPM Insights' dashboards provide a clear action plan for project and portfolio managers and reflects the issues, inconsistencies, or schedule problems that take place on projects at the moment.[37]

PPM Insights is a collection of modules that perform separate functions in support of project management. Two examples of modules are Proactive Risk Management and Flexible Configuration. Proactive Risk Management is a module that makes the dashboard a place where project managers can see notifications about issues, inconsistencies, and "bottlenecks" on their projects, so that they are able to take preventive actions before incurring any negative impact on project delivery.[38] Flexible Configuration is a module that allows project managers to configure flexible dashboards that can be easily modified to suit the organization's project management processes, such as focusing on Key Performance Indicators (KPIs) and other important parameters that define project success.[39]

PPM Insights' algorithms do not just roll up multiple projects into a portfolio they way humans do. It restructures portfolio management processes, and by doing so, it restructures how humans think about project and portfolio management. PPM Insights collects project data from multiple project management platforms and integrates them into portfolio data on a single platform, from which it uses the portfolio data to generate data analytics, forecast future statuses of projects, and produce simulations to support decision-making about enterprise strategy. PPM Insights' algorithms do not simply perform project and portfolio management faster than humans. The algorithms restructure the processes in project and portfolio management. By restructuring project and portfolio management processes, the algorithms have the effect of restructuring the way humans think about them. The human thinking about project management transitions from humans conducting projects individually to humans experiencing an algorithm-induced restructure of thought about project management.

IBM - Watson Studio

International Business Machines Corporation (IBM) built an Artificial Intelligence based technology named Watson Studio to support project management.[40] The architecture of Watson Studio is centered around a project. A project is established as an online collaborative workspace, that is a container where project participants organize resources and work with data.[41] First, Watson Studio creates a workspace for working on a project.[42] Then, Watson Studio uses Natural Language Processing (NLP), which understands human language, to populate the workspace with an integration of disparate documents such as PDFs, Charts, Tables, Call Logs, handwritten documents, News articles, Tweets and information stored in the cloud.[43] Watson Studio can synthesize unstructured data, specific to a particular industry and a particular company.[44]

Watson Studio does not conduct project management activities the way humans do; Watson Studio restructures project management.[45] When combined with the full pipeline of IBM technologies, Watson Studio is capable of automating processes that make up a project's full lifecycle.[46] Watson Studio prepares data in the workspace for collaborative work by data scientists, and domain experts.[47] It also combines project data sets with organizational assets for business purposes.[48] IBM Watson's Machine Learning algorithms enable data scientists, developers and project managers to accelerate deployment by harnessing Artificial Intelligence at scale across multiple platforms.[49]

An artificially intelligent system designed to take a holistic approach to project management, IBM Watson Studio constructs a workspace in which Natural Language Processing (NLP) reads and organizes disparate sources of project information. The algorithms do not simply perform project management faster than humans. The algorithms restructure the way that project management is done. By interacting with IBM Watson Studio, project managers restructure their thinking about project management. The human thinking about project management changed from humans organizing projects individually to humans experiencing an algorithm-induced restructure of thinking about projects.

These technologies, Commit Assistant, PPM Insights and Watson Studio, all contribute evidence for my observation that project management is bifurcating into two roles: data management and strategy management. Data management is about the collection of data, the number-crunching, the data analytics, and the production of dashboards. Strategy management is about the simulations of business operations, and decision making that affects an organization's strategic goals. Increasingly, data management is being performed by technology, while strategy management is attracting project and portfolio managers.

When project managers acquire knowledge and master a skill, their knowledge and skill are offloaded to technology. Artificial Intelligence based project management skills may translate to artificially intelligent systems about organizing projects and assembling projects into portfolios. Machine Learning algorithms mine large volumes of project management data to discern patterns that enable the algorithms to learn how project managers conduct projects. Despite the offload from human project managers to artificially intelligent systems, technology does not conduct project management the way humans conduct project management. Technology restructures the data, restructures the processes, and eventually effects a restructuring of the way humans think about project management. This is in line with Gartner's trend from technology-literate people toward people-literate technology. Both of those trends are in line with Giegerich's trend from the semantical level to the syntactical level of psychology. In the next sections, I explain that the PMI's trend in project management has commonalities with Gartner's trend in technology and Giegerich's trend in psychology. These trends contribute to the overarching paradigm shift, which I believe puts Homo Sapiens on the cusp of generating a new version of humanity.

Comparing PMI's Trend In Project Management With Gartner's Trend In Technology

Gartner's trend in technology supports the PMI's trend in project management. I describe Gartner's trend in detail in Chapter 1: "Gartner's Trend In Technology".

As a reminder, this is how I summarize Gartner's trend in technology:

There is a paradigm shift …

From a model of Technology-Literate People,
where individuals acquire knowledge about digitalization,

To a model of People-Literate Technology,
where algorithms acquire knowledge about people.

In the 20th century, the project management tools commonly used were the Gantt Chart, Critical Chain Project Management (CCPM), and Project Evaluation Review Technique (PERT).[50] Project managers had to become technology-literate about those tools in order to use them in project management. Those tools are built from applications that embody project management knowledge and skills that people offloaded to technology.

In the 21st century, project management tools include Commit Assistant, PPM Insight, Monday.com, and IBM Watson Studio. These tools are built from Machine Learning algorithms that are capable of learning about human project management by mining historic data. Some of those tools are people-literate, in the sense that they learn about people by mining historical data collected from people. Commit Assistant is learning about the detection of errors in computer code written by people who are software developers. PPM Insights is learning about simulation and data analytics by mining data collected from portfolio and project management. IBM Watson Studio is learning how to model data gathered by Natural Language Processing of project information created by project managers using multiple media.

Twenty first century project management technology does not perform project management in the same way that human project managers do. Technology may learn aspects of project management from historical data collected from human project managers, but when digitalizing project management, technology restructures data and processes. To find errors in computer code, developers read the code while analyzing the logic written in the code. Commit Assistant identifies errors in computer code by detecting patterns learned from historical data. People manage projects by carving up requirements into manageable tasks and assigning resources, then carefully balancing adjustments in work, cost and time, while keeping stakeholders updated about progress of the projects. PPM Insights restructures the data of multiple projects into a portfolio on one platform, monitors key parameters for detecting variances that pose risk, generates action plans to mitigate risk, and produces dashboards for simulation of business operations. When project managers collaborate with customers and domain experts, project assets are generally separate from business assets until completion of a project. The pipeline of IBM technologies behind Watson Studio enables the establishment of a collaborative workspace in which project managers, customers and domain experts all collaborate using project assets and business assets throughout the project's lifecycle.

In line with Gartner's trend in technology, Machine Learning systems like Commit Assistant, PPM Insights and Watson Studio are becoming increasingly people-literate. It is noteworthy that they all perform project and portfolio management activities without mimicking humans. Instead of duplicating what humans do, the Machine Learning systems learn from historical project management data, then restructure the ways that project data are organized, restructure the ways that processes are defined and restructure the presentations of project information in support of decision making. The examples of digitalized project management described in this chapter provide evidence that Gartner's trend in technology is consistent with the project management shift from technology-literate people to people-literate technology.

Comparing PMI's Trend In Project Management With Giegerich's Trend In Psychology

The PMI's trend in project management is also consistent with Wolfgang Giegerich's trend in psychology. I describe Giegerich's trend in detail in Chapter 2: "Wolfgang Giegerich's Trend In Psychology".

Here is a reminder of how I summarize Giegerich's trend in psychology:

There is a paradigm shift …

From a focus on the semantical level of psychology,
where individuals engage in the individuation process,
a goal-seeking effort to differentiate their minds from the
unconsciousness of their communities,

To a focus on the syntactical level of psychology,
where human culture engages in the interiorization process,
an intellectual discipline of interpreting phenomena
that emerge in the world.

Giegerich's trend is from the semantical level to the syntactical level of psychology. When people performed project and portfolio management in the 20th century, they were performing at the semantic level of psychology. People had to become literate about technologies like Critical Path Analysis, Gantt Chart Analysis and Project Review Evaluation Technique to be able to engage in formal project management. Those technologies were acquired incrementally. Each added a separate tool to the project management toolkit. These project management tools were developed by individuals differentiating their thinking from the prevailing project management outlook at the time. The incremental addition of project management technologies by individuals differentiating themselves from mainstream project management outlook place 20th century project development at the semantical level of Giegerich's trend in psychology.

Twenty first century project management technologies draw on historical project management data gathered from the activities of many project managers. The technologies combine multiple projects into portfolios that represent the strategic goals of enterprises. Those enterprises that cultivate a digital culture are the ones that the PMI describe as gymnastic enterprises. The gymnastic enterprises interpret the phenomenon of technology differently from the traditional enterprises. Gymnastic enterprises are open to the restructuring that technology brings to project management. The restructuring of data, of processes and of decision making do not just change project management activities; it also restructures the way humans think about project management. By interacting with project management technology, humans restructure their thinking about project management. By their lesser focus on digitalization of project management, traditional enterprises miss out on the restructuring of human thought and remain at the semantical level of psychology. By taking advantage of the restructuring of human thought that comes with the digitalization of project management, gymnastic enterprises place themselves at the syntactical level of psychology. The human thinking about project management transitioned from humans conducting projects individually to humans experiencing an algorithm-induced restructure of thinking about project management. That transition is compatible with Giegerich's trend from the semantical level to the syntactical level of psychology.

Summary

The Project Management Institute's (PMI's) 2021 *"Pulse of the Profession"* reports a trend of gymnastic enterprises emerging from the pool of traditional enterprises as they become more invested in cultivating a digital culture. The PMI's trend in project management is supported by Gartner's trend in technology. In the past, project management was conducted in traditional enterprises, where technology-literate people prevailed. Now, there is a trend toward a new style of project management in an emerging type of enterprise. In my view, PMI's traditional enterprises, that are slow to adopt a digital culture, tend to attract technology-literate people, while the PMI's gymnastic enterprises, that embrace a digital culture, tend to attract people-literate technology. Commit Assistant, PPM Insight, and IBM Watson Studio are examples of people-literate technology that populate gymnastic enterprises.

Gartner's trend moves from technology-literate-people to people-literate-technology. The technology-literate-people style of project management is about people offloading to technology the labor-intensive, number-crunching activities of balancing the relationships among work, resources and cost. Those are tactical activities. What the offloading accomplishes is that it enables project managers to devote their effort to other activities, such as using technology-generated analytics and simulations to carve alternate paths from which to choose the future of their gymnastic enterprises. Those are strategic activities. Machine Learning algorithms in project management tools such as Commit Assistant, PPM Insight, and IBM Watson Studio are facilitating a trend in project management that is supported by Gartner's trend in technology. The transition from project managers becoming literate about 20th century tools, to tools becoming literate about 21st century project managers, is compatible with Gartner's trend from technology-literate people, to people-literate technology.

The PMI's trend in project management is also supported by Giegerich's trend in psychology. In the past, project management involved the incremental acquisition of technical knowledge. That is evident in the way people learn and accumulate knowledge about project management. Critical Path Analysis, Gantt Chart Analysis and Project Review Evaluation Technique are technologies that are acquired incrementally. These project management tools were developed separately and were added to the project management toolkit building block style. Project management that depends on individual consciousness occurs at the semantical level of psychology. Now, there is a trend toward a new style of project management in an emerging type of enterprise. In my view, the trend is not merely about project management moving from a lesser focus to a greater focus on building a digital culture. The trend is also about movement from the individual to the cultural. The project management involved in building a digital culture mines data gathered from a collective mass. For example, Commit Assistance learned to detect errors by mining ten years of errors made by a collective mass of developers. That kind of learning is qualitatively different from how an individual learns. When algorithms learn from mining data gathered from people, the knowledge is usually acquired through pattern detection, and it distills knowledge obtained from a collective mass of people. That knowledge is collective knowledge. It belongs, not to an individual, but to an enterprise. Algorithms learn from human collective knowledge. When humans interact with the algorithms, it has the effect of restructuring human thinking and puts gymnastic enterprises at the syntactical level of psychology.

When algorithms apply the knowledge they gather from patterns in data, they restructure the process of project and portfolio management. At the syntactical level, there is a restructuring of knowledge and process. Knowledge that had been acquired incrementally is restructures to enable it to offer a more integrated holistic service to a culture. Processes that were dependent on incremental knowledge and individual consciousness at the semantical level are restructured for integrated knowledge. At the syntactical level, processes are performed by algorithms and do not necessarily depend on individual human consciousness.

It was Israeli historian Yuval Noah Harari who pointed out that intelligence is being de-coupled from consciousness. While I agree with that, I also see another perspective. I believe that intelligence is bifurcating into human intelligence and artificial intelligence. I see the bifurcation taking place in project management, in those enterprises that the PMI designates gymnastic enterprises. I also believe that the bifurcation of intelligence puts Homo Sapiens on the edge of generating a new version of humanity. The current version of humanity populates the PMI's traditional enterprises, while the new version of humanity is incubating in the PMI's gymnastic enterprises. The new version of humanity is having its thinking restructured as it develops a worldview that is less individual and more cultural.

Commit Assist, PPM Insights and IBM Watson Studio are examples of technology that facilitates the restructuring of thought. These algorithm-based project management tools do not just duplicate project management tasks and perform them faster than humans. Algorithms restructure the performance of project management. As a result of working with project management technology, project managers restructure their own thinking about project management. The PMI's trend in project management is compatible with Giegerich's trend in psychology, because algorithms have the effect of restructuring human thinking, and that places them at the syntactical level of psychology. Together, the PMI's trend in project management, Gartner's trend in technology and Giegerich's trend in psychology, all point to Homo Sapiens being on the brink of generating a new version of humanity.

NOTES:

1. See web site: Project Management Institute, <u>Project Management Institute | PMI</u>, Article "Membership".
2. PMI's *Pulse of the Profession* 2021.
3. *"A Guide to the Project Management Body of Knowledge (PMBOK® Guide)"* Seventh Edition.
4. <u>Project Management Institute.</u>
5. <u>International Project Management Association.</u>
6. See web site: Project Management Institute.Budapest, <u>2021 and beyond.... 5 Disruptive Trends in Project Management | PMI Budapest</u>, Article "2021 and Beyond .... 5 Disruptive Trends in Project Management" by Antonio Nieto-Rodriguez.
7. See web site: CEM Solutions, <u>Project Management History - A Story of Evolution - CEM Solutions</u>, Article "Project Management History – A Story of Evolution".
8. See web site: Project Management.com, <u>ProjectManagement.com - Scaled Agile Framework (SAFe)</u>, Article "Scaled Agile Framework (SAFe)".
9. See web site: Project Management.com, <u>ProjectManagement.com - Scaled Agile Framework (SAFe)</u>, Article "Scaled Agile Framework (SAFe)".
10. See web site: Project Management Institute, <u>Pulse of the Profession (2021) | PMI.</u>
11. See web site: Project Management Institute, <u>Pulse of the Profession (2021) | PMI.</u>
12. See web site: Project Management Institute, <u>Pulse of the Profession (2021) | PMI.</u>
13. See web site: Project Management Institute, <u>Pulse of the Profession (2021) | PMI.</u>
14. See web site: REPLICAN, <u>10 Project Management Trends Emerging in 2021 | Replicon</u>, Article "10 Project Management Trends Emerging in 2021" by Arpan Patra.

15. See web site: REPLICAN, <u>10 Project Management Trends Emerging in 2021 | Replicon</u>, Article "10 Project Management Trends Emerging in 2021" by Arpan Patra.

16. See web site: REPLICAN, <u>10 Project Management Trends Emerging in 2021 | Replicon</u>, Article "10 Project Management Trends Emerging in 2021" by Arpan Patra.

17. See web site: REPLICAN, <u>10 Project Management Trends Emerging in 2021 | Replicon</u>, Article "10 Project Management Trends Emerging in 2021" by Arpan Patra.

18. See web site: REPLICAN, <u>10 Project Management Trends Emerging in 2021 | Replicon</u>, Article "10 Project Management Trends Emerging in 2021" by Arpan Patra.

19. See web site: International Project Management Association (IPMA), <u>The future of project management: Global Outlook 2019 (ipma.world)</u>, Article "The Future of Project Management: Global Outlook 2019" by KPMG, AIPM and IPMA.

20. See web site: AIM (Analytics India Magazine), <u>How Ubisoft Is Mainstreaming Machine Learning Into Game Development (analyticsindiamag.com)</u>, Article "How Ubisoft Is Mainstreaming Machine Learning Into Game Development".

21. See web site: AIM (Analytics India Magazine), <u>How Ubisoft Is Mainstreaming Machine Learning Into Game Development (analyticsindiamag.com)</u>, Article "How Ubisoft Is Mainstreaming Machine Learning Into Game Development".

22. See web site: AIM (Analytics India Magazine), <u>How Ubisoft Is Mainstreaming Machine Learning Into Game Development (analyticsindiamag.com)</u>, Article "How Ubisoft Is Mainstreaming Machine Learning Into Game Development".

23. See web Site: KitGuru, <u>Ubisoft is using machine learning to spot bugs before they make it into the final game code | KitGuru</u>, Article "Ubisoft is using machine learning to spot bugs before they make it into the final game code".

24. See web Site: KitGuru, Ubisoft is using machine learning to spot bugs before they make it into the final game code | KitGuru, Article "Ubisoft is using machine learning to spot bugs before they make it into the final game code".

25. See web Site: KitGuru, Ubisoft is using machine learning to spot bugs before they make it into the final game code | KitGuru, Article "Ubisoft is using machine learning to spot bugs before they make it into the final game code".

26. See web Site: KitGuru, Ubisoft is using machine learning to spot bugs before they make it into the final game code | KitGuru, Article "Ubisoft is using machine learning to spot bugs before they make it into the final game code".

27. See web site: PPM Express, About PPM Express - Portfolio Visibility & Integrated PPM Software, Article "About PPM Express".

28. See web site: PPM Express, About PPM Express - Portfolio Visibility & Integrated PPM Software, Article "About PPM Express".

29. See web site: PPM Express, About PPM Express - Portfolio Visibility & Integrated PPM Software, Article "About PPM Express".

30. See web site: Integrated Project Portfolio Management, Project Portfolio Management Software - Integrated, Secure Cloud PPM, Article "Integrated Project Portfolio Management".

31. See web site: Integrated Project Portfolio Management, Project Portfolio Management Software - Integrated, Secure Cloud PPM, Article "Integrated Project Portfolio Management".

32. See web site: Integrated Project Portfolio Management, Machine Learning for Project Management- ML-Based Portfolio Management - PPM Insights, Article "Machine Learning for Project Management".

33. See web site: Integrated Project Portfolio Management, Machine Learning for Project Management- ML-Based Portfolio Management - PPM Insights, Article "Machine Learning for Project Management".

34. See web site: Integrated Project Portfolio Management, Machine Learning for Project Management- ML-Based Portfolio Management - PPM Insights, Article "Machine Learning for Project Management".

35. See web site: Integrated Project Portfolio Management, Machine Learning for Project Management- ML-Based Portfolio Management - PPM Insights, Article "Machine Learning for Project Management".

36. See web site: Integrated Project Portfolio Management, Machine Learning for Project Management- ML-Based Portfolio Management - PPM Insights, Article "Machine Learning for Project Management".

37. See web site: Integrated Project Portfolio Management, PPM Express - PPM Insights Dashboard - Project Manager's Action Plan - All in One Place, Article "Project Manager's Action Plan".

38. See web site: Integrated Project Portfolio Management, PPM Express - PPM Insights Dashboard - Project Manager's Action Plan - All in One Place, Article "Project Manager's Action Plan".

39. See web site: Integrated Project Portfolio Management, PPM Express - PPM Insights Dashboard - Project Manager's Action Plan - All in One Place, Article "Project Manager's Action Plan".

40. See web site: IBM.com, About Watson | IBM, Article "IBM Watson is AI for Business".

41. See web site: IBM.com, About Watson | IBM, Article "IBM Watson is AI for Business".

42. See web site: IBM.com, About Watson | IBM, Article "IBM Watson is AI for Business".

43. See web site: IBM.com, About Watson | IBM, Article "IBM Watson is AI for Business".

44. See web site: IBM.com, About Watson | IBM, Article "IBM Watson is AI for Business".

45. See web site: IBM.com, About Watson | IBM, Article "IBM Watson is AI for Business".

46. See web site: IBM.com, About Watson | IBM, Article "IBM Watson is AI for Business".

47. See web site: IBM.com, <u>About Watson | IBM</u>, Article "IBM Watson is AI for Business".
48. See web site: IBM.com, <u>About Watson | IBM</u>, Article "IBM Watson is AI for Business".
49. See web site: IBM, <u>Machine Learning | IBM</u>, Article "IBM Watson Machine Learning".
50. See web site: Hive, <u>A Brief History of Project Management | Hive</u>, Article "A Brief History of Project Management" by Michaela Rollings.

SEVEN

Daniel Solove's Trend In Reputation Management

There is a paradigm shift …

*From a social norm where individuals
control their own reputation,*

*To a social norm where algorithms
control people's reputations.*

This chapter is about a trend in reputation management discerned by international expert in privacy law, Daniel Solove. I describe Solove's paradigm shift from a social norm where individuals control the management of their reputations, to a social norm where algorithms control the management of people's reputations. The paradigm shift is toward a situation where Machine Learning algorithms generate reputation scores as measures of people's reputations. There is a proliferation of web sites that produce reports and scores about people's reputations, without involvement or approval of the people whose reputation they score. It appears that for every reputation scoring web site, there is a web site offering to enhance reputations by downplaying the negative and emphasizing positive aspects of reputations. In this chapter, I also explain my view that the control of reputation management is bifurcating into personal control and societal control.

The sources of information that shape my thinking in this chapter are:

- *"The Future of Reputation: Gossip, Rumor, and Privacy on the Internet"* by Daniel J. Solove[1]
- *"A Critical Dictionary of Jungian Analysis"* by Andrew Samuels, Bani Shorter and Fred Plaut[2]
- Web site: ReputationDefender[3]
- Web site: NetReputation[4]
- Web site: COINTRUST[5]
- Web site: crowdspring[6]
- Web site: Reputation Protection Online[7]
- Web site: WebiMax[8]

Daniel Solove's Trend In Reputation

This is how I summarize Solove's trend in reputation management.

There is a paradigm shift ...

*From a social norm where individuals
control their own reputations,*

*To a social norm where algorithms
control people's reputations.*

This is how Solove describes reputation management:[9]

"There's a paradox to the heart of reputation —- despite the fact that we talk about reputation as earned and the product of our behavior and character, it is something given to us by others in the community. Reputation is a core component of our identity —- it reflects who we are and shapes how we interact with others —- yet it is not solely our own creation. ... Our reputation depends upon how other people judge and evaluate us .... Our good reputation can quickly be lost, with deleterious consequences to our friendships, family, jobs and financial well-being. ... The law ... allows people to protect their reputations from being sullied by falsehoods. But ... why should we have a right to control it at all?

Under one theory of reputation, the law professor Robert Post observes, it is a form of property. People earn the esteem of others by 'the fruit of personal exertion'. ... One reason to protect reputation, then, is to preserve the years of effort people put into developing it. Another theory of reputation, Post notes, is that we protect it in the name of human dignity. As Post explains, the 'dignity that defamation law protects is thus respect (and self-respect) that arises from a full membership in a society'.

...

Another reason to protect reputation is that the stakes are so high .... Our reputation matters ... to others in society, who use it to determine whether to trust us. ... When an individual has a better reputation than deserved, it might be the result of essential facts being concealed ... Hence the conflict: we want information to flow openly, for this is essential to a free society, yet we also want to have some control over the information that circulates about us."

Solove's book "*The Future of Reputation*" provides an account of the changes in reputation management facilitated by the Internet. Before the Internet, people managed their own reputations by selecting the personal information they choose to share with others. For example, there is personal control of reputation in resumes submitted to prospective employers and applications submitted for membership in professional associations. Another example is societal control of reputation administered by consumer credit reporting agencies that manage credit scoring to assess credit worthiness of individuals in the marketplace. Since the Internet came into common usage, the management of reputation is being taken over by online organizations that gather information from sources such as social media, government agencies, law enforcement and court records, professional associations and personal memberships in clubs.

Reputation management is not a formally established discipline. Anyone with Internet access can set up a reputation scoring service. Equally, anyone with Internet access can set up a service to correct errors in reports about reputation. In addition, anyone with Internet access can set up a service to enhance reputations by downplaying the negative while emphasizing the positive aspects of a person's life. Organizations that produce reputation scores tend to regard their reputation managing algorithms as propriety information and, as a consequence, they are not inclined to publish the criteria by which their algorithms calculate reputation scores. People whose reputations are scored are not consulted about the accuracy of information being collected about them.

They may not know that they are the subject of reputation reports being published about them to an unlimited Internet audience. Many hiring managers use online reputation scores as a factor in deciding who to hire as employees. Some online dating organizations make reference to online reputation scores in matching partners. Several vendors who offer products and services to the public rely on financial credit scores to decide who to accept as customers.

In writing about the prevalence of information on the Internet, Solove states:[10]

> "The Internet allows information to flow more feely than before. We can communicate and share ideas in unprecedented ways. ... We're heading toward a world where an extensive trail of information fragments about us will forever be preserved on the Internet, displayed instantly in a Google search."

This is how he describes social control and social norms related to reputation:[11]

> "Thus beyond allowing individuals to guard against dealing with dishonest people, reputation also functions to preserve social control. By ensuring that people are accountable for their actions, reputation gives people a strong incentive to conform to social norms and to avoid breaching people's trust."

Here is Solove's comment about information being gathered by technologies:[12]

> "We live in an age when many fragments of information about our lives are being gathered by new technologies, hoarded by companies in databases, and scattered across the Internet."

He points out that reputation is at the core of identity:[13]

"Our reputation can be a key dimension of our self, something that affects the very core of our identity. Beyond its internal influence on our self-conception, our reputation affects our ability to engage in basic activities in society."

Solove also informs us that the law safeguards people's reputation:[14]

"With the rise of the modern economy, honor ceased to be the core of a person's reputation, ... the vehicle for people to safeguard their reputations, ... the courts became the main option. ... the law of privacy and the law of defamation."

This is how he points out that credit reporting is an example of technology measuring reputations of people:[15]

"Credit reporting agencies and other companies provide heaps of data about individuals for employer background checks."

For an unspecified amount of time, the Internet preserves multiple sources of personal communication such as blogs, online discussion groups, chatrooms, YouTube postings, and Facebook postings. Similarly, the Internet stores public records like births, deaths, marriages, divorces, levels of education, professional associations, financial credit scores, club memberships, employers, self-employment, patents, community awards. A Google search reveals personal communications and public records to anyone who has Internet access.

The personal and public information are not all accurate, but the Internet is the source upon which employers, business partners, friends, strangers, dates, neighbors and relatives rely. An authority on privacy law, Solove sees the uncontrolled flow on information on the Internet as an online collision between free speech and privacy. He provides examples to show that the unfettered flow of information on the Internet can impede opportunities for self-development and freedom. Although the Internet is generally considered a facilitator of free speech, traditional ideas of reputation management need to be revised to accommodate a balance between privacy and free speech.

Where Solove sees a collision between free speech and privacy, I see a bifurcation occurring in the control of reputation management. Reputation scoring organizations do their scoring independently of the individuals whose reputation they score. As individuals try to exercise personal control of their reputations, they have to contend with reputation management organizations publishing scores which, accurate or not, are being relied upon by employers, vendors, dating partners, business colleagues, relatives and friends.

The bifurcation is about a split in control of reputation management. Those who own the technology to generate reputation scores have unfettered control over reputation management, while individuals have limited control over their own personal information. This bifurcation matters, not just because prospective employers, vendors and friends rely on reputation scores determined by algorithms, but also because reputation is an important part of one's identity. One's sense of identity guides the path one carves out in life. It also influences interaction with others. There is an important relationship between reputation and persona. Reputation is about how others view an individual. Persona is about how an individual presents themself to the world.

In ""A Critical Dictionary of Jungian Analysis" psychologists Andrew Samuels, Bani Shorter and Fred Plaut provide the following definition of "persona":[16]

> **Persona:** "The term derives from the Latin word for mask worn by actors in classical times. Hence, persona refers to the mask or face a person puts on to confront the world. Persona can refer to gender identity, a stage of development (such as adolescence), a social status, a job or a profession. Over a lifetime, many personas will be worn and several may be combined at any one moment.

> Jung's conception is ... that there is an inevitability and ubiquity

to the persona. In any society, a means of facilitating relationship and exchange is required; this function is partly carried out by the personas of the individuals involved. ...

It follows that persona is not to be thought of as inherently pathological or false. There is a risk of pathology if a person identifies too closely with his/her persona. This would imply a lack of awareness of much beyond social role (lawyer, analyst, laborer), or gender role (mother), and also a failure to take account of maturation (for instance, an apparent failure to adapt to having grown up). Persona identification leads to a form of psychological rigidity or brittleness; the unconscious will tend to erupt into consciousness rather than emerging in a manageable way. The ego, when it is identified with the persona, is capable only of external orientation. It is blind to the internal events and hence unable to respond to them. It follows that it is possible to remain unconscious of one's persona.

These last comments point to the place Jung assigns to persona in the structure of the psyche. That was as mediator between the ego and the external world."

As the mediator between ego and external world, the persona faces the world of online reputation management organizations offering a variety of scores ... reputation score, credit score, creativity score ... which are determined independently of the person being scored, and for which there is little visibility into the scoring criteria. That can set up a dissonance between an individual's persona and reputation. An unfettered use of information gathered from multiple sources on the Internet can disrupt a person's identity formation. In my opinion, the tension between societal control of reputation and personal control of reputation is creating a bifurcation in the control of reputation management. If societal control of reputation overwhelms personal control of reputation, it can interfere with the person's development of self-awareness. If the persona one presents to the world is too far removed from the person within, that creates instability and may inhibit opportunities for self-development.

The bifurcation in control of reputation brought to public attention the proliferation of reputation management organizations that do not disclose the online sources from which they select information, nor the criteria for proprietary algorithm scoring. There are no means of determining whether the score produced by one organization is more reliable than another. There is no way to find out if a score has been improved by a reputation enhancement organization, nor what was improved.

There is no consistency among the assortment of organizations that produce and enhance reputation scores. People who afford to have their reputations enhanced will, unfairly, have better scores. People who do not monitor their online presence, will not notice if their online reputations are inaccurate. Hiring managers, social clubs, professional associations do not have a reliable source of reputation reporting. The many discrepancies between people's personas and their online reputations are driving a need for a more consistent and reliable way for people to discover each other's reputations.

Solove sees the trend in reputation management moving toward a situation where society composes a code of ethics for reputation management. As a privacy lawyer, he regards failure to comply with the code of ethics as justification for recourse to legal action. I agree with Solove that a society determined code of ethics would add transparency and accountability to online reputation management.

What follows are examples of web based organizations that use algorithms for reputation management.

Reputation Protection Online – calculates reputation scores

Reputation Protection Online is a service that uses algorithms to produce reputation scores for individuals and companies,

offers information about the factors that influence the calcula-
tion of reputation scores, and helps with improvement of the
digital footprints that determine reputation scores.[17] In addition
to providing reputation scores online, Reputation Protection
makes recommendations about strategies to improve reputation
scores and protect reputations.[18] Reputation Protection indicates that their reputation scores fall in the following ranges:[19]

- **Good reputation score:** 80% and above
- **Average reputation score:** More than 50% but less than 80%
- **Poor reputation score:** Less than 50%

The Reputation Protection Online web site does not indicate whether a
score of exactly 50% is regarded as average or poor.

This information is useful because reputation scores can have significant
impacts on individuals as well as companies. Here are some of the possible impacts:[20]

- Individuals who apply for jobs can be rejected because of poor
  reputation score.

  The skills of an individual may match the job descriptions of
  open positions, but the individual may be rejected due to a
  poor reputation score.

- Banks can refuse to provide services because of poor reputation
  score.

  Banks can check reputation scores before deciding whether
  to provide banking services such as opening an account, ap-
  proving loans or credit card applications.

- Credit Card accounts can be cancelled because of poor reputation score.

Processing of applications for credit cards involve screening an applicant's online presence.

- Personal relationships can be affected by poor reputation score.

People check reputation on the Internet before starting professional and personal relationships.

- Companies can lose market share due to poor reputation scores.

Consumers tend to favor companies that have good online reputations because of the risk that companies with low reputation scores may not survive in the marketplace.

WebiMax – repairs online reputations

WebiMax is an online organization that uses algorithms to help clients build a positive digital reputation by assigning a dedicated project manager, who arranges the following services:[21]

- Conduct a free in-depth analysis of the client's existing online reputation reports.
- Set realistic goals for enhancing the client's reputation.
- Apply cost-effective techniques to bury negative content and promote positive content in the client's reputation report.
- Apply techniques to downplay negative content, and delete where possible.
- Determine which positive content to promote, and rank them.
- Populate the first page of Google search results with positive content, since most users do not scroll beyond the first page,
- Regularly monitor the client's online presence.
- Continuously update WebiMax algorithms to match changes in Google algorithms.

PAID Network Protocol – scores reputation in compliance with contracts

PAID Network Protocol generates legal toolkits that use algorithms to provide a decentralized business infrastructure which accommodates reputation scoring.[22] PAID Network Protocol provides users with a legal toolkit of proprietary templates for setting up legally binding contracts without the need for expensive legal services.[23] Their blockchain based reputation scoring algorithms use biometric signatures built into a mechanism for ensuring that the parties involved in a contract get paid, and for eliminating users who repeatedly violate the terms of contractual agreements, as indicated by decreasing reputation scores.[24] PAID Network Protocol allows users to complete the business agreement process from beginning to end through the platform, enabling the PAID community to participate in lucrative business sectors which have historically been accessible only to the largest worldwide institutions.[25]

The PAID marketplace allows businesses and professionals to post offers or requests for various types of business agreements. After an expression of interest in an offer or a request, PAID Network Protocol provides users easy-to-use web and mobile apps, enabling them to do business with anyone, anywhere, without the need for expensive legal services.[26] PAID Network scores the reputation of users by use of Decentralized Digital Identities (DID) for its reputation scoring system.[27] When a new user creates a PAID profile, their identity is verified via DID which includes biometric factors, and their reputation score is set at zero. Depending on feedback obtained from people in the PAID community, the reputation score will decrease or increase. Users who comply with the terms of agreement in their contracts get positive feedback, and so increase their reputation scores. Those who do not comply get negative feedback and their reputation scores decrease.[28] A PAID user's behaviors are transparent to other users in the PAID community. Reputation scores allow users to evaluate trustworthiness of prospective business partners in the PAID community, based on records of adherence to terms of agreement in business transactions. Since PAID's user identification depends on biometric authentication, a user with a low reputation score cannot simply start a new PAID user profile. Duplicate biometric signatures are not admissible to the PAID community.[29]

Crowdspring – calculates reputation scores for creative artists

Crowdspring has clients who post projects that require creative work, and invites individuals to submit designs for the projects. When an individual first submits a design for a project, crowdspring measures the quality of work and calculates a reputation score on a scale of 0 to 100.[30] Crowdspring's algorithm adjusts the reputation score periodically taking into account client feedback, and project awards among other factors that are not disclosed by crowdspring.[31] Crowdspring's algorithms produce reputation scores in the following ranges:[32]

- **Highest reputation score:** Greater than 90
- **Above average reputation score:** Between 80 and 90
- **Average reputation score:** Between 70 and 80
- **Poor reputation score:** Less than 70

The crowdspring web site does not indicate the category for a score that is exactly 70, or exactly 80, or exactly 90 points.

Crowdspring invites clients to post projects for which they will pay creative artists. Client feedback influences the reputation scores assigned to creative artists. When a new creative artist joins the crowdspring community, crowdspring assigns a provisional reputation score of 70.[33] The reputation score is later adjusted based on reviews from clients and awards for projects, among other factors.[34]

ReputationDefender – suppresses negative content & creates positive content

ReputationDefender is an online organization that offers algorithm based services of reputation protection to customers including private individuals, small businesses, executives, and large corporations.[35] The services are about controlling online reputations by protection of privacy and correction of misleading search results. One service is a free reputation report card for customers to find out how others see them online.[36] Another service is control of information that displays when there is a Google search for the customer.[37] ReputationDefender offers a privacy service that prevents third-party companies from selling a customer's sensitive personal, or corporate, information online.[38]

As part of the effort to improve online reputations, Reputation Defender encourages their customers to recruit loyal associates to write online reviews that have positive content.[39] In addition, ReputationDefender offers corporate cybersecurity to reduce the risk of social engineering attacks against a corporation's personnel.[40] This is the process that ReputationDefender uses to curate information about online reputation:[41]

- **Strategy:** Develop an engagement strategy specifically for improving a customer's reputation
- **Content:** Create positive content about the customer
- **Publication:** Strategically publish content about the customer online
- **Suppression:** Inhibit dissemination of negative search results about the customer, and
- **Control:** Expertly manage the process for optimal results about the customer's reputation.

> *ReputationDefender is lobbying Congress to give Americans "the right to remove old, inaccurate, or misleading items from their personal search results".[42]*

Reputation Protection Online, WebiMax, PAID Network Protocol, crowdspring and ReputationDefender all offer reputation management services in their own particular style. They lack the coherence that would enable them to provide a reliable service to society. For example, they have different categories for reptation scoring. Reputation Protection Online has three categories for scoring reputation: good, average and poor. Crowdspring has four categories: highest, above average, average and poor. A score in one organization has a different meaning from the same score in another organization. For example, Reputation Protection Online regards a 60% score as average, while crowdspring regards a 60% score as poor.

Reputation management organizations produce reports about people's reputation; but they have no industry standards and they do not publish the scoring criteria in their proprietary algorithms. As a consequence of there being no formal mechanism for removing inaccurate, old or misleading content in reputation reports, a niche opened up for other organizations to provide a service of making corrections to inaccurate reputation reports. That created another niche for still other organizations to offer services to enhance reputations by suppressing negative content and promoting positive content.

The evolution of reputation management raises questions. How do reputation management organizations distinguish between gossip and fact when collecting data from social media? How are hiring managers to distinguish between initial reputation reports and enhanced reports? When one organization scores a reputation as average and another organization scores the same reputation as poor, how can anyone rely on reputation scores?

Comparing Solove's Trend In Reputation Management With Gartner's Trend In Technology

Gartner's trend in technology is discernable in reputation management. Daniel Solove's trend in reputation management demonstrates a technological underpinning in which algorithms are becoming literate about people's reputations. I describe Gartner's trend in detail in Chapter 1: "Gartner's Trend In Technology".

As a reminder, this is how I summarize Gartner's trend in technology:

There is a paradigm shift ...

From a model of Technology-Literate People,
where individuals acquire knowledge about digitalization,

To a model of People-Literate Technology,
where algorithms acquire knowledge about people.

Gartner's trend in technology provides technical underpinning for Solove's trend in reputation management. Solove's trend goes from an era of technology-literate people who curate their own reputations, to an era of people-literate technology when algorithms curate people's reputation. In the first decade of the 21st century, professionals learned to use technology, such as LinkedIn, to manage their professional reputations. As the Internet came into wide use, a profusion of web-based organizations began offering reputation management services. Organizations such as Reputation Protection Online, WebiMax, PAID Network Protocol, crowdspring and ReputationDefender all offer reputation management services that are based on algorithms which acquire literacy about people, by mining online sources of data about people. As the volume of data about people increases, the algorithms are becoming increasingly literate about people.

Algorithms mine multiple sources of online data to calculate reputation scores and produce profiles about people. Because there are so many organizations producing inconsistent information about reputations, the technological trend in reputation management does not appear to be evolving. For monetary gain, technology is being used to splinter reputation management into fragmented marketing services that involve the following activities:

- Select reputation related information from public and private sources available online
- Produce reputation reports for entities such as people, companies, creative work and Internet domains
- Calculate reputation scores
- Correct inaccuracies in reputation reports, and
- Enhance reputations by suppressing negative information and emphasizing positive information.

Although technology adds more dimensions to reputation management, technology is not being used to provide society with a reliable service. There is hardly any visibility into criteria used in algorithms, and providers of reputation management services accept little accountability for their products. Society cannot rely upon the results due to the ad hoc approach to visibility, accountability and reliability of reputation management services.

Comparing of Solove's Trend In Reputation Management With Giegerich's Trend In Psychology

Wolfgang Giegerich's trend in psychology is not yet discernable in reputation management. Daniel Solove's paradigm shift in reputation management is not aligned with Wolfgang Giegerich's paradigm shift in psychology. I describe Giegerich's paradigm shift in detail in Chapter 2: "Giegerich's Trend In Psychology".

This is how I summarize Giegerich's trend in psychology:

There is a paradigm shift ...

From a focus on the semantical level of psychology,
where individuals engage in the individuation process,
a goal-seeking effort to differentiate their minds from the
unconsciousness of their communities,

To a focus on the syntactical level of psychology,
where human culture engages in the interiorization process,
an intellectual discipline of interpreting phenomena
that emerge in the world.

I found no indication that Solove's trend in reputation management has anything in common with Giegerich's trend in psychology. The reputation management industry is currently too fractured to provide a useful service to humanity. In Solove's view, the reputation management industry blurs the distinction between privacy of information and freedom of speech, and diminishes the separation of gossip from facts, because the industry lacks a society-determined code of ethics. In my opinion, the online reputation management organizations have fragmented the industry into functions that lack the coherence which would enable the provision of a reliable service. Although the reputation management industry is well supported by algorithms that are becoming literate about people, their effect is one of disruption without any movement away from the semantical level of psychology.

The shift from personal to impersonal, from individual to cultural, is not yet noticeable in reputation management. Due to the lack of a shift from an individual level to a cultural level, consciousness has achieved neither a new type of awareness, nor a new scope of action. Reputation management is still at the individual level, where technological progress occurs in an unreflective, unconscious manner, and where meaning is trapped in the physical nature of technology. It is unreflective because there is little regard for the meaning of technological progress for humanity in reputation management. It is unconscious because the meaning of technological progress in reputation management remains implicit.

The psyche is challenged to release the trapped meaning into consciousness and reveal the explicit meaning of technological progress in reputation management. At the cultural level, the psyche has not yet become conscious of the trapped meaning of digitalized reputation management. Humans have just started to reflect on the objective process in which they are engaged in advancing technology in reputation management. Daniel Solove is among those promoting a cultural purposiveness behind digitalization of reputation management. He expects the cultural purposiveness to be determined by a society defined code of ethics, backed up by a legal recourse when reputation is tarnished. I expect the trapped meanings of changing identity, shifting control of reputation management, and a sustainable tension between reputation and persona to become conscious. The challenge is for consciousness to think more abstractly about technological reputation management, and in doing so, raise itself from the personal level to the impersonal level, in other words, from the semantical to syntactical level of psychology.

Before the Internet, reputation was an attribute of the individual, determined by interaction between individual and the community. Individuals differentiate themselves from others in their community by carving out a unique path in life based on their aspirations, their accomplishments, and their outlook on life. They accumulate educational credentials, prizes for athletic achievements, awards for community services, professional credentials. All of which are achieved independently. They conduct their lives according to the values in their belief systems. Individually, they conduct their reputation management by presenting selected information about themselves to employers, clubs, professional associations, political parties. What is revealed and what is concealed are up to the individual. This approach to reputation management places humans at the semantical level of psychology.

With the availability of the Internet and multiple online sources of personal information, Western culture is demanding more from reputation management. Enabled by Machine Learning algorithms, Western culture is interpreting reputation as a phenomenon that encompasses more than just individuals. There are reputation scoring algorithms for any entity for which information can be digitized, for example, people, corporations, creative work, and Internet domains. The reputation management industry has responded by providing multiple incremental services, without recognition of the need for an integrated service that can be useful to the collective whole that is representative of Western culture. So far, the disruption of reputation management has not been productive. Instead of human knowledge of reputation management being restructured, it has become fragmented. In reputation management, the site of agency is shifting from individuals, but has not yet found its place in the collective whole that pervades human culture. That approach keeps the reputation management industry tied to the semantical level of psychology. The reputation management industry does not yet demonstrate the coherent agency that would place it among the industries functioning at the syntactical level of psychology.

Summary

The reputation management industry produces reputation profiles and scores that can make or break the academic lives, social lives or careers of individuals. While wielding the power to render an individual's life a success or a failure, the reputation management industry has created a situation where people have the enormous responsibility to regularly monitor their online presence and make corrections to reputation reports that are published about them, without their involvement or permission. The onus is on individuals to discover the existence of multiple online reputation management organizations and correct any inaccuracies they report. That is not a sustainable arrangement. Society is demanding that the reputation management industry be held accountable for disclosing criteria by which algorithms calculate reputation scores, and be held responsible for making corrections of errors in their data gathering and reputation scoring.

Reputation management has the potential to become a useful service to society. Human interaction requires a reliable source of information about reputations. Technology is available to support the digitalization of reputation management. Homo Sapiens has the resources and the capabilities, but does not yet have the will to overhaul the management of reputation. Bifurcation of the control of reputation management into societal control and personal control is underway, but the resulting dissatisfaction has not yet reached the level of chaos that warrants a new approach.

What Solove's work contributes to the overarching paradigm shift is the awareness of the trend in social norm regarding reputation management, from control by individuals to control by algorithms. Where Gartner's trend underpins Solove's trend is in the growth of Machine Learning technology that enables algorithms in organizations like ReputationDefender, WebiMax and NetReputation to acquire knowledge about people. Hiring managers, professionals, dating partners, and business colleagues all depend on being able to check reputations online before making decisions about interactions with others. However, the technology has fragmented reputation management into splinters that do not serve society well. Giegerich's trend in psychology is not evident in reputation management. The reputation management concepts and activities that Solove describes are all at the semantical level of psychology, where consciousness of reputation management increases incrementally, but there is no restructuring of outlook, and by implication, no restructuring of consciousness.

Although Solove's trend in reputation management has not attained the kind of service to humanity that would place it on Giegerich's syntactical level of psychology, a bifurcation in control of reputation management points to a coming disruption that will give rise to a new outlook. In addition to the bifurcation in control of reputation management, there is also a splintering in types of services: reporting reputation, correcting reputation, and enhancing reputation. Moreover, the subject of reputation management is fragmenting into multiple entities, for example, person, company, creativity, and Internet domain. I agree with Solove's recommendation that what remains for digital reputation management to become a more reliable service is for society to develop a code of ethics. I expect the code of ethics to address topics such as: Which organizations are eligible to report on reputation? What are acceptable sources of public and private information for inclusion in reputation reporting? Who determines the criteria for reputation scoring? How will the public get visibility into the criteria? What is the basis for determining that a reputation should be corrected? Should the enhancement of reputations be allowed? Who is accountable for errors in reputation reports? When society sanctions a code of ethics, reputation management will be among the areas of life contributing to Homo Sapiens' creation of a new version of humanity.

NOTES:

1. See "*The Future of Reputation: Gossip, Rumor, and Privacy on the Internet*" by Daniel J. Solove, Yale University Press, 2008.
2. See "*A Critical Dictionary of Jungian Analysis*" by Andrew Samuels, Bani Shorter and Fred Plaut, Routledge & Kegan Paul Ltd., 1987.
3. See web site: ReputationDefender, ReputationDefender | Online Reputation Management.
4. See web site: NetReputation, Online Reputation Management Agency for Businesses and Individuals :) (netreputation.com).
5. See web site: COINTRUST, CoinTrust.com – About Us - CoinTrust.
6. See web site: crowdspring, What should creatives know about the new reputation score? | Help | crowdspring.
7. See web site: Reputation Protection Online, Reputation protection online – Take Charge of your reputation before anyone takes it for granted.
8. See web site: WebiMax, Online Reputation Management Company | WebiMax.

9. See *"The Future of Reputation: Gossip, Rumor, and Privacy on the Internet"* pp 58-61.

10. See *"The Future of Reputation: Gossip, Rumor, and Privacy on the Internet"* p 27.

11. See *"The Future of Reputation: Gossip, Rumor, and Privacy on the Internet"* p 54.

12. See *"The Future of Reputation: Gossip, Rumor, and Privacy on the Internet"* p 17.

13. See *"The Future of Reputation: Gossip, Rumor, and Privacy on the Internet"* p 52.

14. See *"The Future of Reputation: Gossip, Rumor, and Privacy on the Internet"* pp 219 - 220.

15. See *"The Future of Reputation: Gossip, Rumor, and Privacy on the Internet"* p 56.

16. See *"A Critical Dictionary of Jungian Analysis"* p 107 – 108.

17. See web site: Reputation Protection Online, <u>Check Reputation Score for free – Reputation protection online</u>, Article "Why You Need to Know Your Reputation Score".

18. See web site: Reputation Protection Online, <u>Check Reputation Score for free – Reputation protection online</u>, Article "Why You Need to Know Your Reputation Score".

19. See web site: Reputation Protection Online, <u>Check Reputation Score for free – Reputation protection online</u>, Article "Check Online Reputation".

20. See web site: Reputation Protection Online, <u>Check Reputation Score for free – Reputation protection online</u>, Article "Check Online Reputation".

21. See web site: WebiMax, <u>Reputation Management & Repair Services (webimax.com)</u>, Article "We can fix and protect your online presence fast".

22. See web site: COINTRUST, <u>PAID Network Protocol Face Network Attack, Hacker Gains $3.16mln Worth Ethereum (cointrust.com)</u>, Article "PAID Network Protocol Face Network Attack, Hacker Gains $3.16mln Worth Ethereum".

23. See web site: Law Insider, <u>Reputation Scoring Sample Clauses | Law Insider</u>, Article "Reputation Scoring Sample Clauses".
24. Se web site: Law Insider, <u>Reputation Scoring Sample Clauses | Law Insider</u>, Article "Reputation Scoring Sample Clauses".
25. See web site: PAID, <u>Support : PAID Network</u>, Article "Launching only the top tier projects in the space".
26. See web site: PAID, <u>Support : PAID Network</u>, Article "Launching only the top tier projects in the space".
27. See web site: PAID, <u>Support : PAID Network</u>, Article "Launching only the top tier projects in the space".
28. See web site: PAID, <u>Support : PAID Network</u>, Article "Launching only the top tier projects in the space".
29. See web site: PAID, <u>Support : PAID Network</u>, Article "Launching only the top tier projects in the space".
30. See web site: crowdspring, <u>What should creatives know about the new reputation score? | Help | crowdspring</u>, Article "What should creatives know about the new reputation score".
31. See web site: crowdspring, <u>What should creatives know about the new reputation score? | Help | crowdspring</u>, Article "What should creatives know about the new reputation score".
32. See web site: crowdspring, <u>What should creatives know about the new reputation score? | Help | crowdspring</u>, Article "What should creatives know about the new reputation score".
33. See web site: crowdspring, <u>What should creatives know about the new reputation score? | Help | crowdspring</u>, Article "What should creatives know about the new reputation score".
34. See web site: crowdspring, <u>What should creatives know about the new reputation score? | Help | crowdspring</u>, Article "What should creatives know about the new reputation score".
35. See web site: ReputationDefender, <u>ReputationDefender | Online Reputation Management</u>, Article "Solutions as unique as your reputation".
36. See web site: ReputationDefender, <u>ReputationDefender | Online Reputation Management</u>, Article "Solutions as unique as your reputation".

37. See web site: ReputationDefender, <u>ReputationDefender | Online Reputation Management</u>, Article "Solutions as unique as your reputation".

38. See web site: ReputationDefender, <u>ReputationDefender | Online Reputation Management</u>, Article "Solutions as unique as your reputation".

39. See web site: ReputationDefender, <u>ReputationDefender | Online Reputation Management</u>, Article "Solutions as unique as your reputation".

40. See web site: ReputationDefender, <u>ReputationDefender | Online Reputation Management</u>, Article "Solutions as unique as your reputation".

41. See web site: ReputationDefender, <u>ReputationDefender | Online Reputation Management</u>, Article "Solutions as unique as your reputation".

42. See web site: ReputationDefender, <u>ReputationDefender | Online Reputation Management</u>, Article "Solutions as unique as your reputation".

EIGHT

Robert Lanza's Trend
In Biocentrism

There is a paradigm shift ...

*From a focus on consciousness as an attribute of
individuals who are detached observers of reality,*

*To a focus on consciousness as an attribute of collective
groups who are active co-creators of reality.*

In this chapter, I explain how Robert Lanza's paradigm shift in biocentrism demonstrates a bifurcation in the attribute of consciousness. I also explain how Robert Lanza's trend in biocentrism parallels Gartner's trend in technology, as well as Wolfgang Giegerich's trend in psychology. Biocentrism makes a distinction between the consciousness of an individual and the consciousness of a collective group, in terms of how they relate to physical reality. In the context of biocentrism, Lanza reports a trend that is moving from a focus on consciousness as an attribute of an individual who observes physical reality with detachment, to a focus on consciousness as an attribute of a collective group which actively co-creates physical reality.

Robert Lanza is a research scientist in biology. He collaborates with theoretical physicist Matej Pavsic and astronomer Bob Berman in offering a new perspective on the role of consciousness in science. They outline principles of biocentrism that indicate consciousness is not merely about individuals observing reality with detachment, but also about collective groups of people who are actively co-creating reality.

My sources of information for this chapter are:

- *"A Grand Biocentric Design: How Life Creates Reality"* co-authored by Robert Lanza, Matej Pavsic and Bob Berman [1]
- *"ONE WORLD: The Health and Survival of the Human Species in the 21st Century"* edited by Robert Lanza [2]

This is my characterization of the paradigm shift that Lanza notices in science:[3]

There is a paradigm shift ...

From a focus on consciousness as an attribute of individuals who are detached observers of reality,

*To a focus on consciousness as an attribute of collective
groups who are active co-creators of reality.*

Of course, consciousness continues to be an attribute of both individuals and collective groups. What is shifting is our focus. In this chapter, I describe Lanza's paradigm shift in terms of examples of consciousness as an attribute of individuals, as well as examples of consciousness as an attribute of collective groups. In those centuries that preceded the 21st century, when our focus on consciousness was primarily regarded as an attribute of individuals, two examples of detached observers of physical reality were:

- Leonardo Da Vinci who produced the drawing of Vitruvian Man, and
- Nicolaus Copernicus who published a hypothesis of heliocentricity.

In the 21st century, when our focus on consciousness is shifting to an attribute of collective groups, here are two examples of collective groups who are co-creating reality:

- Doctors Without Borders are co-creating emergency aid for disasters around the world, and
- Urban developers are co-creating Smart Cities on digital foundations in countries worldwide.

Lanza points out that science is acknowledging the role of consciousness in the creation of reality:

"… (S)tarting around a century ago, physics took an abrupt turn and began to seriously consider that without consciousness, the material universe alone could not supply the true or complete picture of reality."[4]

He also informs us that science gives consciousness a new definition:

"… (C)onsciousness is, as it turns out, a quantum phenomenon."[5]

This is how Lanza describes the role of consciousness as being central to the paradigm shift in science:[6]

"…(A)ctivities at the subconscious level are in a quantum superimposition – meaning, all possibilities simultaneously coexist. But the moment their results pop into reality and conscious awareness, a perceptible 'choice' is made. This is key because there are always many possible chains of brain activity … But when consciousness hangs up on one of them – subjectively perceived as the awareness of a definite outcome – this can be mathematically described as the collapse of the wave function.

…

Wave function collapse is triggered by the individual's perception of one particular possibility over the other possibilities."

Historically, consciousness has been defined in different ways in different domains. For example, in neuroscience, consciousness is considered a biological phenomenon and a byproduct of neuronal activity in the brain.[7] Some schools of psychology define consciousness as an emergent phenomenon, that is, a quality possessed by a human, but not possessed by any of the components that make up a human.[8] Neuroscience does not explain how the brain produces consciousness. Neither does biology. In physics, particles appear to respond to a conscious observer. While consciousness is acknowledged by scientists, many dismiss it as not being relevant to the physical world. Robert Lanza explains that their view misses out on the findings that it is consciousness, not matter, which plays the primary role of creating reality. The observer is a prerequisite for physical reality to "materialize" from probability waves of energy.

What Lanza offers goes beyond the mere addition of another definition of consciousness. He offers the notion that consciousness is central to a paradigm shift about how we understand reality. We used to view physical reality as detached observers. Now, biocentrism informs us that we are not merely detached observers; we are also co-creators of physical reality. In the domain of biocentrism, consciousness is regarded as a quantum phenomenon. That is, in making a choice among many possibilities, consciousness brings about the collapse of a wave function of possibilities.[9] The brain processes information by enabling us to collapse a probability wave of possibilities into a single experience.[10] The brain's involvement in the collapse of a wave into a singular experience occurs as the brain processes information through electrical and chemical signals.[11]

Lanza's explanation is that:

> " ... (I)t is modulation of the ion dynamics at the quantum level that allows all parts of the information system that we associate with consciousness – with the unitary 'me' feeling – to be simultaneously interconnected." [12]

Biocentrism is the context in which Lanza describes the paradigm shift about how we regard consciousness. Biocentrism is about the relationship between the universe and consciousness. In the remainder of this chapter, I summarize the principles of biocentrism, then I focus on those principles that relate to the paradigm shift in consciousness. Biocentrism stands in opposition to anthropocentrism, which views the world in terms of individual human experiences. Biocentrism views the world in terms of collective human experiences. Anthropocentrism ascribes consciousness to the individual, while biocentrism ascribes consciousness to humanity as a whole.

Principles of Biocentrism

In defining the principles of biocentrism, Lanza makes reference to an information system. He associates consciousness with an information system. That information system does not appear in the index of his book

"A Grand Biocentric Design: How Life Creates Reality" and it is not defined in Chapter 16 where he lists the principles of biocentrism. I interpret his use of the expression "information system" based on my reading of his Chapter 6: "Consciousness" and Chapter 7: "How Consciousness Works".[13] Lanza seems to use the expression "information system" as a metaphor for the mechanism by which consciousness weaves together all the components input from multiple sources in order to sustain an awareness of an identity and participate in the co-creation of reality. The input components include sensory experiences from seeing, touching, hearing, smelling and tasting. In addition, there are non-rational components that arise from a state of reverie, when one is neither asleep nor fully awake, such as intuition. Other non-rational components come from memories and emotions. A rational component is thinking that derives from deduction and induction. Then, there are components that come from the forces that shape the dimensions of the physical world: nuclear, electromagnetic and gravitational forces.

In Lanza's information system, consciousness weaves together input components from those sources to navigate a path through life in the physical world. During that navigation, consciousness uses the mental construct of time to throw awareness backward when remembering the past, to maintain awareness of the present, and to throw awareness forward when planning for the future. Consciousness also uses the mental construct of space as cognitive architecture for activities such as understanding and representing the layout of physical environments, organizing objects, interacting with others and discerning spatiotemporal relationships in the physical world. With that interpretation of information system in mind, here is my summary of the principles of biocentrism as laid out in Lanza's book:[14]

- **First principle of biocentrism**: Our perception of reality is an ongoing process that involves our consciousness using the mental constructs of space and time as tools of the human mind.
- **Second principle of biocentrism**: Our perceptions of the external physical world and our internal mental world are bound

together and cannot be separated.

- **Third principle of biocentrism**: The behavior of objects in the external physical world is tied to the presence of an observer. Without an observer, they would exist in an undetermined state of probability waves.
- **Fourth principle of biocentrism**: Without the consciousness of an observer, "matter" exists in an undetermined state of probability. If any universe existed before consciousness, it could only have existed in a probability state.
- **Fifth principle of biocentrism**: Biocentrism is useful in explaining the structure of the universe because humankind creates the universe, rather than the other way around. The structure of the universe is the result of humans applying the spatiotemporal logic of our minds.
- **Sixth principle of biocentrism**: Time is a mental construct; it does not have any physical existence outside of human perception. The flow of time is the concept by which we perceive change in the universe.
- **Seventh principle of biocentrism**: Space is a mental construct; it has no physical existence outside of human perception. The three dimensions of space form a concept that enables humans to understand the physical world about us.
- **Eighth principle of biocentrism**: Biocentrism provides an interpretation of how mind, matter and consciousness are interconnected by the way they all feed into the same information system.
- **Ninth principle of biocentrism**: The human mind constructs reality by reference to forces (nuclear, electromagnetic, gravitational) that have their basis in the components of the information system. Each force indicates how energy interacts at different levels of forces (weak forces, strong forces).
- **Tenth principle of biocentrism**: To reconcile the two major topics of physics (quantum mechanics and general relativity), it is necessary to take observers into account.
- **Eleventh principle of biocentrism**: The structure of physical

reality is defined by observers.

In the domain of biocentrism, Lanza defines consciousness as a quantum phenomenon. He explains the quantum phenomenon by pointing out that at the subconscious level, many possibilities exist as a wave function, which collapses into a single experience when triggered by consciousness focusing on one possibility.[15]

The paradigm shift that Lanza discerns starts with a focus on consciousness being an attribute of individuals, then shifts to a focus on consciousness being noticed as an attribute of collective groups. I identify two individuals who lived prior to the 21st century, when consciousness was mainly regarded as an attribute of individuals. These examples are Leonardo Da Vinci and Nicolaus Copernicus. Each man used his individual consciousness and mental constructs of space and time to describe aspects of the physical world, at a time when people used to become technology-literate in order to acquire knowledge about the physical world. Both individuals proposed a structure of an external aspect of the physical world based on how they directed their consciousness to acquire the technology-literacy necessary to their observation of physical reality. I also describe two examples of collective groups that currently exist in the 21st century as functioning institutions. They are Doctors Without Borders and urban developers of Smart Cities. Each is a collective group using its consciousness to co-create an evolving institution in the physical world, at a time when there exists people-literate technology which provides digital foundations for the evolving institutions.

The following are descriptions of Leonardo Da Vinci's "Vitruvian Man" drawing and Nicolaus Copernicus' hypothesis of heliocentricity.

Consciousness As An Attribute Of An Individual: Leonardo Da Vinci

Leonardo Da Vinci was an Italian who lived in the period from 1452 to 1519. He was designated a "Renaissance Man" to indicate that he was among those humans who believed that man exists at the center of the universe, with unlimited potential for development and the responsibility to acquire knowledge for realizing that potential.[16] Da Vinci developed his interests in the anatomy of the human body, sculpture, humanism, painting and mathematics. One of his more famous works is the "Vitruvian Man" drawing. The Vitruvian drawing did not originate with Da Vinci. His inspiration came from an architectural guide written by a Roman architect, who made the original drawing between 30 and 15 BC.[17] The Vitruvius drawing was intended to combine the geometry of architecture of buildings with the geometry of the ideal male body for the purpose of constructing buildings to house humans.[18]

Combining his interests in art and mathematics, Da Vinci elaborated the Vitruvian drawing to portray the ideal proportions of a male human body. The drawing superimposes images of an ideal nude man in different positions all within a circle: a standing position with legs together, standing with legs apart, standing with arms outstretched, and standing with arms raised. The superimposition of multiple poses enabled Da Vinci to determine several proportions of the ideal male body.[19] Here are some of the detached observations that Da Vinci made about measurements of the ideal male body:[20]

- The length of a man's outspread arms is equal to his height.
- If a man spreads his legs enough to reduce his height by one-fourteenth, and raises his arms until the middle fingers touch the level of the top of his head, the center of the outspread limbs will be in the navel, and the space between the legs will be an equilateral triangle.
- The distance from the roots of his hair to the bottom of his chin is one tenth of a man's height.
- The distance from the bottom of his chin to the top of his head is one eighth of a man's height.

- The distance from the top of his breast to the top of his head is one sixth of a man's height.
- The distance from the top of his breast to the roots of his hair is one seventh of a man's height.
- The distance from his nipples to the top of the head is one fourth of a man's height.
- The greatest width of the shoulders is one fourth of a man's height.
- The distance from elbow to tip of the hand is one fifth of a man's height.
- The distance from elbow to the angle of the armpit is one eighth of a man's height.
- The length of his whole hand is one tenth of a man's height.
- The length of his foot is one seventh of a man's height.
- The beginning of his genitals is the middle of the man.
- The distance between the bottom of his chin and his nose is equal to the distance between the roots of his hair and his eyebrows.

In Da Vinci's lifetime, architects and sculptors observed the physical world by becoming technology-literate about the tools that were available to them. Da Vinci acquired the technology-literacy to produce the drawing while he worked as an apprentice studying architectural and technological design in the workshop of Andrea del Verrochio.[21] The technology available to Da Vinci included pen, ink, metalpoint stylus and paper. He learned how to use them to sketch images.[22] Metalpoint stylus is a silver-tipped instrument used in a particular technique of drawing on specially prepared paper. A metalpoint stylus drawing is made by inserting a thin metal rod into a holder and applying it to a prepared surface. Generally, the preparation involves coating textured paper of medium thickness, or a wood panel. As the stylus moves across the textured surface, it deposits small pieces of metal and leaves a legible mark on the surface.[23] The coating is usually made up of a mixture of calcium carbonate to give body, a pigment to provide color, and a glue that holds all the ingredients together. The resulting mixture is then brushed or rolled onto paper and allowed to dry.[24]

Fifteenth century architects believed that the circle and the square have symbolic powers. Circles were associated with the 'cosmic and the divine' while squares were associated with the 'earth and the secular'.[25] Da Vinci's insertion of the ideal masculine form into a frame made of a circle on top of a square implied two intentions. One intention was to determine the proportions of the masculine body. The second intention was to demonstrate the perfection of the universe by showing how humans fit into the earth and the cosmos.[26] Da Vinci practised his skill by spending long periods sketching anything he observed outdoors. His *Vitruvian Man* is an example of how Da Vinci applied consciousness, as an attribute of an individual, in detached observation of physical reality, as expressed in the form of the human body. In producing what is now a famous image in the Western world, he acquired technology-literacy about the geometry of architecture and applied it to capture the proportions of the human body.

Architect Vitruvius and Renaissance Man Da Vinci did not collaborate on the drawing; they were not contemporaries. Each worked as an individual. Architect Vitruvius applied his consciousness to the relationship between architecture and the masculine body. Da Vinci applied his consciousness to observing, measuring and calculating the proportions of the ideal male body. The fifteenth century drawing of Vitruvian Man is the product of Da Vinci using his mental construct of space to create a model of the male human body. This is an example of an individual using his consciousness as a detached observer of the physical world, before Lanza's paradigm shift. The Vitruvius drawing is an example of consciousness being used as an attribute of an individual, first by the architect Vitruvius, then decades later, enhanced by Da Vinci.

Consciousness As An Attribute Of An Individual: Nicolaus Copernicus

Nicolaus Copernicus was a Polish astronomer, mathematician and Catholic canon law expert, who lived from 1473 to 1543.[27] He was called a polymath, meaning that he was knowledgeable about a wide range of topics and able to apply his knowledge to resolve issues. In Copernicus' lifetime, astronomers made observations about the physical world by becoming technology-literate about the tools at their disposal. Copernicus acquired the technology-literacy to propose the heliocentric model by learning to use the technology available to him. In his hometown Frombork in Poland, Copernicus learned how to use an observatory to track the orbits, conjunctions and eclipses of Mercury, Mars, Venus, Jupiter and Saturn.[28]

Copernicus credited Greek and Arabic astronomers with the origin of the heliocentric model. During Copernicus' lifetime, it was Ptolemy's geocentric model of planetary movement that prevailed as the accepted model, but Copernicus disputed certain points that differed from his own observation of planetary orbits. Here are some of the detached observations that Copernicus made about planetary motions:[29]

- The Earth is one of several planets revolving around a stationary sun in a determined order.
- The Earth has three motions: daily rotation, annual revolution, and annual tilting of its axis.
- Retrograde motion of the planets is explained by the Earth's motion.
- The distance from the Earth to the Sun is small compared to the distance from the Sun to the stars.

The technology available to Copernicus included the Tusi couple, trigonometry, the equinoxes of solar theory, the Julian calendar, planetary longitude and latitude, and an observatory he built for himself in Frombork in 1513.[30] The Tusi couple is a mathematical device that Copernicus used to convert the circular motion of planets to linear motion. The Julian calendar was known to have shortcomings because it did not correspond to the length of the solar year. Pope Leo X promoted the need for calendar reform, but Copernicus, in his role as Catholic canon law expert, advised the pope that astronomical theory needed to be corrected before calendar reform. However, Copernicus did not publish his heliocentric model until the end of his life. One reason could be that his model was so counterintuitive that he was afraid of possible ridicule from other astronomers and the public. Another possibility is that the heliocentric model ran counter to the prevailing Catholic biblical point of view and that he might lose his job in the Catholic Church, or worse be imprisoned for heresy. That second possibility was borne out by the fact that, after Copernicus died, the Catholic Church put Italian astronomer Galileo Galilei under house arrest for promoting Copernicus' heliocentric model. While under house arrest, Galileo used his time to build telescopes which provided evidence that supported Copernicus' hypothesis of heliocentricity.

Da Vinci and Copernicus learned the technologies available in their lifetimes, then applied their technology-literacy as detached observers recording their observations about physical reality. Da Vinci used his technology-literacy to observe the physical proportions of the human body and sketch his observations in the *Vitruvian Man* drawing. Copernicus applied his technology-literacy to observe the planets in motion and define the heliocentric model. In the 15th and 16th centuries, the lives of Da Vinci and Copernicus provide examples of situations where the notion of consciousness is exhibited as an attribute of individuals who were detached observers of reality.

In the remainder of this chapter, I offer two examples of the paradigm shift to a focus on consciousness as an attribute of collective groups.

They are Doctors Without Borders and urban developers of Smart Cities. Both are examples of collective groups that are currently using people-literate technology to actively co-create reality in the 21st century. The groups are co-creating reality, supported by technology that is becoming increasingly people-literate:

- **Doctors Without Borders** is a collective group of health-care professionals and logistics experts who are co-creating a global emergency response system supported by technology that becomes increasingly people-literate by learning from data collected about people recovering from disasters.
- **Urban Developers of Smart Cities** are professionals including urban planners and engineers who are co-creating cities supported by technology that becomes increasingly people-literate by learning from data collected about the populations of modern cities.

Consciousness As An Attribute of A Collective Group: Doctors Without Borders

The precursor to Doctors Without Borders was an organization with the French name Medecins San Frontiers (MSF), which started in late 20th century when six doctors responded to the Red Cross appeal for medical volunteers to help victims in the war-torn Nigerian province of Biafra.[31] MSF was created on the belief that all people have the right to medical care regardless of gender, race, religion, creed, or political affiliation, and that the needs of these people outweigh respect for national boundaries.[32]

The MSF also provided emergency aid after an earthquake in the Nicaraguan capital of Managua, flooding in Honduras after the hurricane Fiji, and to Cambodians during the oppressive rule of the Pol Pot regime. In these early missions, the weaknesses of the MSF as a humanitarian organization became obvious. Preparation for disasters was inadequate, and doctors did not have enough support. At its Annual

General Assembly in 1979, the MSF decided to become more organized. They pooled their efforts in establishing a foundation for a new type of humanitarianism that would give top priority to the wellbeing of victims of disaster, while ignoring political and geographic boundaries. Medecins San Frontiers defined their mission in terms of activity intended to:[33]

- Deliver emergency medical aid to those who need it most.
- Comply with medical ethics to provide care without causing harm.
- Commit to operate with independence, neutrality and impartiality.

They redefined emergency aid by setting up a new type of humanitarianism for helping victims of disasters, such as wars, earthquakes and hurricanes. The MSF also defined new types medicine for delivering humanitarian support:

- War surgery
- Triage medicine
- Public health in the wake of disaster, and
- Education about health-related matters.

To practise these types of medicine, the MSF had to go to the disaster zone where the victims are located. So, they had to overcome national borders. It was helpful to have black and white TV broadcasts show the world the results of war. Blockading forces allowed people to die of hunger and civilians to be killed. The MSF delivered emergency aid and was vocal about the complicity of governments. In 1999, the Medecins San Frontiers was awarded the Nobel Peace Prize.

In the 21st century, the French Medecins San Frontiers became an international organization known as Doctors Without Borders (DWB). DWB operates in over 65 countries around the world. It is an international organization of over 42,000 doctors, nurses, and logisticians. A worldwide collection of people who volunteer to use their medical knowledge and organizational skills to help people in disaster zones. In doing so, they are co-creating a reality of worldwide humanitarian aid by providing the following services:[34]

- They provide medical services across national borders.
- They deliver emergency medical aid to people in disasters zones without consideration of race, creed or politics.
- Their commitment to independence, impartiality, and neutrality reserves for them the power to evaluate medical needs, to access populations without restriction, and to control the distribution of aid without submitting to the priorities of politicians, funders, or the news media.
- They comply with the medical ethics of doing no harm.
- While providing humanitarian support, they give visibility to violations of human rights.
- They are a non-profit organization, whose financial support comes from their fund-raising efforts and from public donations.

Doctors Without Borders (DWB) is a collective group whose consciousness is at a level that is beyond ego consciousness, and whose interests go beyond the interests of individuals. The group is made up of volunteers who have a shared interest in providing service to those in need of care. Functioning as a collective group, Doctors Without Borders erodes geographic boundaries, overrides nationalistic interests, is blind to race, and submits to neither religious doctrine nor political ideology. Doctors Without Borders answers to a global need for health care in areas affected by disasters. This global service puts Doctors Without Borders at Giegerich's level of syntactical psychology, where the focus is on consciousness as an attribute of collective groups. When there is a disaster, Doctors Without Borders dispatches a group of doctors, nurses, logisticians, technicians and other volunteers, all of whom function as a cohesive entity attending to the needs of those affected by the disaster. The sophistication of the technology used to support humanitarian aid places Doctors Without Borders at the higher end of Gartner's technological trend, where technology is becoming literate about people.

Data about emergency aid and humanitarian activities are continuously being collected and stored in artificially intelligent systems. Algorithms built into the artificially intelligent systems are capable of Machine Learning, that is, accumulating knowledge about people in disaster settings, by mining large volumes of data that are continuously being harvested. The technology is becoming increasingly people-literate as more data are harvested from a variety of disaster settings. The knowledge acquired about people recovering from disasters is being used to improve humanitarian activities and prepare for future disasters. There are several digital tools available for supporting humanitarian aid, including ImpactMapper, MedShare and NetHope.[35] IBM partners with Doctors Without Borders in building INTERSOS, a tool for responding to global refugee crises.[36] Here, I choose Telemedicine as an example of an artificially intelligent system that supports the activities of Doctors Without Borders and is becoming people-literate.

Telemedicine engages a network of medical specialists whose expertise is accessible to doctors who are working in situations of emergency humanitarian aid and need verification or advice on diagnoses while in the field of a disaster.[37] The people in Doctors Without Borders are volunteers who are experienced clinicians, medical specialists, logisticians, technicians and administrators. Some work in the field, others work remotely, and they are connected through technology to a digital response space created by Telemedicine. Telemedicine is modern technology that makes it possible to render health care when the provider and the patient are not located in the same place.[38] Telemedicine gives doctors, specialists and patients access to Health Insurance Portability and Accountability Act (HIPAA) compliant video-conferencing tools.[39] While working in the field, doctors may not have easy access to specialists to help them address unusual cases that are outside of their medical experience. Telemedicine enables doctors to record images and text about a patient for electronic transmission to remotely located specialists who make suggestions for diagnosis and treatment. Because of the fast speed of electronic transmission, a response can often be obtained while the patient is still in the doctor's office in the field and available for treatment. Types of Telemedicine services include the following:[40]

- **Interactive medicine** is a Telemedicine feature that provides a digital mechanism for doctors and patients to communicate in real-time. Audio and video communications are encrypted before being transmitted across a geographic network to ensure compliance with Health Insurance Portability and Accountability Act (HIPAA) requirements for security of health information.
- **Store-and-forward** is a Telemedicine feature that allows doctors to keep track of data about cases, so that work which is already done on cases can be viewed by a doctor or specialist who is not on site of the emergency aid. This feature enables doctors to leverage knowledge gained from past diagnoses and treatments.
- **Remote patient monitoring** is a Telemedicine feature that enables doctors to monitor patients remotely and make adjustments to their health care where necessary.

Telemedicine eliminates the need to transfer patients from one hospital to another to be examined by specialists, and for follow-up visits. Telemedicine leverages the health care professionals available in a disaster by extending a team of local professionals to include professionals who serve remotely via electronic media. Telemedicine's storage of electronic medical records creates a bank of data about the practice of medicine in areas where there have been disasters. By mining banks of data about various types of disasters, Telemedicine is becoming people-literate about disasters. It stores information about the involvement of patients, family doctors, nurses and medical specialists. This storage includes information about how medical professionals conduct triage in the context of a disaster, symptoms the patients present, diagnoses made by doctors, treatments administered by nurses, expertise contributed by specialists, patients' response to treatments, which diseases appear in which geographic locations, and which demographic segments are predisposed to which diseases. All of this constitutes an ever-increasing people-literacy that can be retrieved and leveraged for dealing with future disasters. Telemedicine is an example of people-literate technology, which is at the high end of Gartner's paradigm shift in technology.

I describe Telemedicine to show an example of a collective group moving in a scientific trend toward co-creating reality in humanitarian aid. The expression "co-creating" applies because Doctors Without Borders functions as a collective group motivated and directed, not by individual interest, but by collective interest. Doctors Without Borders functions as a cohesive entity that serves humanity without being limited by the boundaries that often limit individual interest: national boundaries, po-litical divisions, family identities, racial heritage and religious beliefs. A humanitarian service is coming into physical reality through the co-cre-ation of the collective group known as Doctors Without Borders.

Consciousness As An Attribute Of A Collective Group: Urban Developers Of Smart Cities

In the 20th century, professionals with particular expertise used to build cities by maintaining essential services as separate domains such as energy management, water management, sanitation, transportation and health care. In the 21st century, professionals are forming collective groups that build Smart Cities by integrating essential services into lay-ers of digital support.[41] There is a paradigm shift away from the view that cities are physical structures and that our perception of them comes from the consciousness of individuals who are detached observers of reality. Traditional mental constructs of cities take the form of a set of co-located residential, commercial and cultural architectural structures supported by services of electricity, water, transportation and security all delivered by different professionals in separate utility companies. The trend is shifting toward a view that cities are developing realities being co-created by collective groups through their shared mental constructs of space, time and urban functionality. Contemporary mental constructs of cities take the form of Smart Cities whose functions are integrated domains of services arranged in layers of digital infrastructure.[42]

There were Smart City research projects, proposals and proof-of-concept initiatives in the first decade of the 21st century. In the year 2011, fifty countries participated in the first Smart City Expo World Congress, which was held in Barcelona.[43]

Here is a list of ten cities considered to be pioneers in the development of Smart Cities:[44]

1. Singapore, Singapore
2. Dubai, United Arab Emirates
3. Oslo, Norway
4. Copenhagen, Denmark
5. Boston, United States of America
6. Amsterdam, Netherlands
7. New York, United States of America
8. London, United Kingdom
9. Barcelona, Spain, and
10. Hong Kong, China.

These cities leverage the technology of the Internet of Things (IoT) in establishing integrated services for residents:[45] Technology companies, urban planners, engineers, elected city officials, public relations promotors and residents all collaborate in setting up Smart Cities. According to a report by McKinsey, the development of a Smart City depends on three layers of computer applications:[46]

- **Data-gathering computer applications:** The first layer is a digital foundation, where data-gathering computer applications combine a critical mass of electronic devices, such as smartphones and sensors, that are all integrated by high-speed communication networks.
- **Data-analyzing computer applications:** A second layer is made up of data-analyzing computer applications that process the incoming streams of raw data from electronic devices to

produce notifications, send alerts, and initiate actions.

- **Information-generating computer applications:** In the third layer, information-generating computer applications are directed at educating residents about the open access to digital resources at their disposal and providing them with up-to-date information about the eight domains that affect the quality of city life.

These are the eight domains of urban services that are known to affect quality of life in Smart Cities:[47]

1. Transportation in the city
2. Healthcare services
3. Security of residents and city infrastructure
4. Water management
5. Energy management
6. Engagement and community services
7. Economic development and housing, and
8. Waste management.

I choose Geolitica as an example of technology that supports infrastructure and security of Smart Cities.[48]

Geolitica is an abbreviation of "geographical analytics" which involves conducting risk assessments using historical crime data to identify the highest-risk locations of specified crimes by geographic location, time of day and day of week. This gives police agencies control over where and when to allocate police officers on daily beats. In addition, Geolitica provides information about when and where extra patrol is necessary for special events like concerts or sporting events. Geolitica maps out geographic patrol boxes of 500 feet squares for the convenience of command staff who determine what locations to designate as patrol areas. Geolitica has a feature for daily communication. Geolitica's platform is used to communicate daily missions and patrol guidance to officers, also to create management reports for crime analysts and command staff.

The values that police agencies derive from Geolitica are accountability and transparency, both of which are high priorities for residents of Smart Cities. Geolitica is about patrol operations management. It gives the residents of Smart Cities transparency into activities of their police department. Geolitica also allows police officers to be accountability to the community of residents in a Smart City. Overall, Geolitica gives police departments opportunities to be effective by allocating police resources in ways that keep Smart Cities safe. The sources of data used by Geolitica are:[49]

- "Records management systems (RMS) databases for crime data.
- Computer assisted dispatch (CAD) databases for collision and selected public safety event data.
- Automated vehicle location (AVL) databases for officer location data."

Headquartered in California, Geolitica services police departments, sheriff's offices, and security companies around the United States as well as internationally.[50] Geolitica was founded in 2012 with the goal of bringing greater transparency and accountability to policing through the use of objective data.[51] Geolitica operates a cloud-based platform that delivers patrol guidance and measures officer performance in real time. Geolitica uses data from departmental records management systems (RMSs) as well as computer-aided dispatch systems (CADs) to prepare and deliver missions and patrol recommendations.[52] Patrol guidance is delivered to any Internet-connected device as a set of 500 x 500-foot boxes on a Google Maps interface. Geolitica uses automatic vehicle location (AVL) to monitor and measure real-time and historical performance of officers during their shifts.[53] All of the data and analytics are available through a set of accountability and transparency reports that can be used internally in police offices and shared externally with elected officials and residents of Smart Cities.

The current Geolitica platform represents a significant investment of over 70 research-years of PhD-level analysis, modeling and development.

Geolitica has delivered over 200 million hours of officer guidance in departments of many sizes around the world.[54] The main features of Geolitica are:[55]

- **Historical data about crime:** Geolitica collects data from police department records management systems to build an artificially intelligent system of complex databases containing historical data about crimes according to types of crime, geographical location of the crimes and dates when the crimes were committed in a Smart City.
- **Movement of police officers during their shifts:** Geolitica also collects historical shift data about the movements of individual officers during their shifts. The purpose is to provide police accountability to the citizenry of a Smart City.
- **Geographic patrol boxes:** Geolitica has a patrol guidance feature, by which it delivers to any Internet-connected device as a set of 500 x 500-foot patrol boxes on a Google Maps interface. Police officers use these patrol boxes to identify areas where there is a high incidence of crimes, so that police departments can allocate resources accordingly. Police also use the patrol boxes to record upcoming events like concerts, political rallies and protest marches.
- **Patrol heatmaps:** Geolitica makes reports available to police departments, for example, patrol heatmaps that show the amounts of time that police officers spend in cities broken down by day and shift. Police use this information to determine if they are over-patrolling or under-patrolling their jurisdictions.
- **Mission reporting about public safety strategy:** Geolitica provides mission reporting to give community leaders opportunity to buy into the police department's public safety strategy for their Smart City.
- **Trends of performance across shifts:** Geolitica uses historical data to produce long-term trends of performance data across shift, beat and command roles. As crime rates go down, police officers become less reactive and more proactive.

Some cities are further along in developing these services than others. Smart Cities in Asia and Europe are more advanced than other continents. All Smart Cities use the Internet of Things for data gathering.[56] Countries where Smart Cities are being established include:[57]

- China
- Denmark
- Netherlands
- Norway
- Singapore
- Spain
- United Kingdom, and
- United States of America.

Integrated city-dwelling services are coming into physical reality through the co-creation of collective groups of urban developers who co-create Smart Cities.

Comparing Lanza's Trend In Biocentrism With Gartner's Trend In Technology

Biocentrism makes note of a qualitative difference between individuals who engage in detached observation of physical reality and collective groups who co-create physical reality. Here I compare Lanza's trend in biocentrism with Gartner's trend in technology. Gartner's trend in technology supports Lanza's trend in biocentrism. I describe Gartner's trend in detail in Chapter 1: "Gartner's Trend In Technology".

As a reminder, this is how I summarize Gartner's trend in technology:

There is a paradigm shift …

From a model of Technology-Literate People,
where individuals acquire knowledge about digitalization,

To a model of People-Literate Technology,
where algorithms acquire knowledge about people.

In comparing Lanza's trend in biocentrism with Gartner's trend in technology, I point out how human agency is shifting from individuals to collective groups, through the support of technology. Renaissance Man Leonardo Da Vinci and polymath Nicolaus Copernicus lived when both trends of biocentrism and technology were focused on the individual. They conducted their lives as detached observers of physical reality. Da Vinci observed the dimensions of human anatomy in the context of architectural design. To do so, he had to learn the technology of architectural design as an apprentice, then study the use of metalpoint stylus for the purpose of drawing *Vitruvian Man*. Copernicus was a detached observer of the orbits of planets in the sky. To generate a heliocentric model, he had to learn how to build and use an observatory. Da Vinci and Copernicus were technology-literate people in their lifetimes. The outcomes of their detached observations were products that humanity continues to value centuries later.

The early part of Lanza's trend in biocentrism is about the focus of attention being on individuals who are detached observers of reality. This parallels the early part of Gartner's trend in technology which is about individuals being technology-literate. During the lifetimes of Da Vinci and Copernicus, Gartner's model of technology-literate people prevailed. Human agency belonged to individuals in the early trends of biocentrism and technology. Individuals demonstrated their agency by learning the technology necessary to meet their goals while observing physical reality with detachment.

Before the establishment of Doctors Without Borders, doctors and nurses functioned as individuals who had earned professional qualifications that allowed them to practice in specific countries. If a doctor wanted to practise medicine in a country other than the country where he or she acquired their qualifications, they would likely have to submit to examinations in the destination country. When a doctor sets up practice, it is usually to offer expertise in a particular branch of medicine, as an individual, in a particular country. Now that Doctors Without Borders is established, with offices in countries all over the world, health care professionals are able to cross borders that separate countries to provide emergency aid as a collective group. They deliver humanitarian aid with strong support from technology, such as Telemedicine, whose people-literacy enables a collective group to provide high-speed service in a digital response space.

Doctors Without Borders and Smart Cities were established in the 21st century, when the trends of biocentrism and technology began to shift focus from individuals to collective groups. A collective group of volunteer doctors co-created Doctors Without Borders. Their mission is to provide humanitarian aid in the aftermath of disasters, without being limited by national borders, racial heritage, religious beliefs or political ideology. Collective groups of urban developers are co-creating Smart Cities. The goal of urban developers is to create higher quality of life for city dwellers by drawing on the collective intelligence of stakeholders including city residents, elected officials, providers of utilities, police command staff and city engineers. To satisfy the goals of co-creation, collective groups in Doctors Without Borders and urban developers of Smart Cities rely on technology that is capable of learning about people. The features of artificially intelligent systems Telemedicine and Geolitica are in line with Gartner's technological trend that is moving toward a model of people-literate technology. Telemedicine is becoming people-literate about recovery from disasters. Geolitica is becoming people-literate about what people do to secure the safety of their cities.

The later part of Lanza's trend in biocentrism highlights the focus of attention on collective groups who are co-creating reality. This aligns to Gartner's model of people-literate technology. In the 21st century, technology is able to continuously gather data about people while mining the data to become people-literate. In the era of Doctors Without Borders and urban developers of Smart Cities, Gartner's model of people-literate algorithms is becoming prevalent. Human agency is shifting from individuals to collective groups in the later trends of biocentrism and technology. Increasingly, collective groups are demonstrating the agency to act in concert for the purpose of achieving their goals, while people-literate technology plays a supporting role.

Comparing Lanza's Trend in Biocentrism With Giegerich's Trend In Psychology

What follows is a comparison of Lanza's trend in biocentrism with Giegerich's trend in psychology. I describe Giegerich's trend in detail in Chapter 2: "Wolfgang Giegerich's Trend In Psychology".

Here is a reminder of how I summarize Giegerich's trend in psychology:

There is a paradigm shift …

From a focus on the semantical level of psychology,
where individuals engage in the individuation process,
a goal-seeking effort to differentiate their minds from the
unconsciousness of their communities,

To a focus on the syntactical level of psychology,
where human culture engages in the interiorization process,
an intellectual discipline of interpreting phenomena
that emerge in the world.

Lanza's trend from a focus on consciousness of individuals to a focus on consciousness of collective groups corresponds to Giegerich's trend from a semantical level to a syntactical level of psychology. In the early part of Lanza's trend in biocentrism, the focus on consciousness of individuals is consistent with the early part of Giegerich's trend, where ego consciousness is dominant and the semantic level of psychology is the prevailing status. In the later part of Lanza's trend in biocentrism, the focus on collective groups corresponds to the later part of Giegerich's trend, where psychological consciousness is dominant and the syntactical level of psychology is the prevailing status.

To compare Lanza's trend in biocentrism with Giegerich's trend in psychology, I point out how human agency is shifting from individuals to collective groups. At the semantical level of psychology, Da Vinci and Copernicus' ego consciousness directed actions that result in incremental additions of content and meaning to knowledge, and by implication to consciousness. What Da Vinci's *Vitruvian Man* drawing added to consciousness were the proportions of the male human body. Those proportions enhanced architectural designs to comfortably accommodate people in architectural structures. What heliocentricity added to consciousness in Copernicus' generation was the opportunity to replace the Julian calendar with the more accurate Gregorian calendar. It was not until decades later that humanity acquired enough scientific knowledge to appreciate the full significance of heliocentricity. The lifetimes of Leonardo Da Vinci and Nicolaus Copernicus both spanned the 15th and 16th centuries. At that time, human agency belonged to individuals. Individuals demonstrated their agency by applying the ego consciousness necessary to meet their goals while observing physical reality with detachment. They functioned at the semantic level of psychology, where knowledge is acquired by degrees, and by implication, consciousness is expanded incrementally.

The later part of Lanza's trend points to attention being focused on collective groups who are co-creating reality. This is consistent with the later part of Giegerich's trend about people functioning at the syntactical level of psychology, where psychological consciousness prevails. At the syntactical level, Doctors Without Borders and urban developers of Smart Cities demonstrate psychological consciousness. The restructure of consciousness generated by Doctors Without Borders is a new way of regarding fellow-humans, as in a borderless, shared existence. The responsibility for helping people recover from disasters changed from being the responsibility of home nations, to becoming a collective responsibility shared across the globe. The effort is spearheaded by volunteer doctors all over the world, while people anywhere in the world can participate by making donations via the Internet. A restructure of consciousness was necessary to co-create Doctors Without Borders and Smart Cities. The restructure of consciousness generated by urban developers of Smart Cities is seen in the redefinition of city dwelling as a collective construction whose stakeholders act in concert to share benefits and responsibilities. The institution of Doctors Without Borders is being co-created by doctors who act in concert to deliver humanitarian aid wherever there are disasters. The practice of establishing Smart Cities is underway in many countries and city dwellers benefit from the availability of algorithms that make it possible, not just to integrate utilities on a digital infrastructure, but also integrate shared consciousness of what constitutes city dwelling. Human agency is shifting from individuals to collective groups in the later trends of biocentrism and psychology. Increasingly, collective groups are demonstrating their agency by the effort that results in restructure of consciousness as they act in concert while co-creating new institutions in physical reality.

Summary

This chapter shows some correspondence among Robert Lanza's trend in biocentrism, Gartner's trend in technology and Wolfgang Giegerich's trend in psychology. All three trends begin with a focus on individuals and progress toward a focus on collective groups. All the trends began before the 21st century and are actively progressing to a higher level of human functionality in the 21st century. The trends in biocentrism, technology and psychology all seem to be discipline-specific aspects of the same overarching paradigm shift, that is, a movement from development of individuals to development of collective groups. In each discipline, artificially intelligent systems play a strong supporting role of enabling collective groups to integrate various expertise into coherent entities that work effectively in providing valuable services to humanity. Technologies like Telemedicine and Geolitica provide the digital infrastructure that is necessary for the establishment of Doctors Without Borders and Smart Cities.

The mindset of individuals like Da Vinci and Copernicus was to acquire the knowledge necessary to observe physical reality with detachment. That mindset put them on the semantical level of psychology. Collective groups such as Doctors Without Borders and urban developers of Smart Cities have a different mindset. By building two institutions that provide services to humanity across the globe, they demonstrate that they are co-creators of physical reality. Doctors Without Borders is a not-for-profit organization whose mission is to provide humanitarian aid worldwide without regard for national borders, race, religion and ideology. That approach restructures the common outlook that the world is divided into national boundaries with national authorities. By implication, the collective group that organizes Doctors Without Borders had to have a restructure of consciousness in order to create a new type of organization that blurs national borders. Similarly, urban developers of Smart Cities had to restructure their outlook in order to establish Smart Cities that are structurally and functionally different from traditional cities. The ability to replace traditional outlooks with radically different outlooks on structure and function of organizations puts these collective groups on the syntactical level of psychology.

What Lanza's work contributes to the overarching paradigm shift is awareness of the cultural sophistication of collective groups in establishing radically different institutions that provide Homo Sapiens with more humanitarian services than traditional organizations. Together, the trends in biocentrism, technology and psychology contribute to the overarching paradigm shift that puts Homo Sapiens on a path to generating a new version of humanity.

NOTES:

1. See "*A Grand Biocentric Design: How Life Creates Reality*" by Robert Lanza, Matej Pavsic and Bob Berman, published by BenBella Books, 2020.

2. See "*ONE WORLD: The Health and Survival of the Human Species in the 21st Century*" edited by Robert Lanza, published by Health Press, 1996.
3. See "*A Grand Biocentric Design: How Life Creates Reality*" by Robert Lanza, pp 75 – 98.
4. See "*A Grand Biocentric Design: How Life Creates Reality*" by Robert Lanza, p 79.
5. See "*A Grand Biocentric Design: How Life Creates Reality*" by Robert Lanza, p 86.
6. See "*A Grand Biocentric Design: How Life Creates Reality*" by Robert Lanza, pp 88 - 89.
7. See web site <u>John Searle on Consciousness and the Brain at TEDxCERN (Transcript) : The Singju Post</u> for a John Searle's description of consciousness as a byproduct of the brain.
8. See web site <u>Frontiers | Phenomenal Consciousness and Emergence: Eliminating the Explanatory Gap | Psychology (frontiersin.org)</u> for a description of consciousness as an emergent phenomenon.
9. See "*A Grand Biocentric Design: How Life Creates Reality*" by Robert Lanza, p 86.
10. See "*A Grand Biocentric Design: How Life Creates Reality*" by Robert Lanza, p 90.
11. See "*A Grand Biocentric Design: How Life Creates Reality*" by Robert Lanza, pp 90 – 91.
12. See "*A Grand Biocentric Design: How Life Creates Reality*" by Robert Lanza, p 91.
13. See "*A Grand Biocentric Design: How Life Creates Reality*" by Robert Lanza, pp 90 - 91.
14. See 12 See "*A Grand Biocentric Design: How Life Creates Reality*" by Robert Lanza, pp 195 – 197.
15. See "*A Grand Biocentric Design: How Life Creates Reality*" by Robert Lanza, pp 86 - 89.
16. See web site <u>Renaissance man | Definition, Characteristics, & Examples | Britannica</u> for definition of Renaissance Man.

17. See web site <u>The Significance of Leonard da Vinci's Famous "Vitruvian Man" Drawing (mymodernmet.com)</u>, for article "The Significance of Leonardo Da Vinci's Famous 'Vitruvian Man' Drawing" by <u>Kelly Richman-Abdou, dated</u> August 5, 2018.

18. See web site <u>The Significance of Leonard da Vinci's Famous "Vitruvian Man" Drawing (mymodernmet.com)</u>, for article "The Significance of Leonardo Da Vinci's Famous 'Vitruvian Man' Drawing" by <u>Kelly Richman-Abdou, dated</u> August 5, 2018.

19. See web site <u>The Significance of Leonard da Vinci's Famous "Vitruvian Man" Drawing (mymodernmet.com)</u>, for article "The Significance of Leonardo Da Vinci's Famous 'Vitruvian Man' Drawing" by <u>Kelly Richman-Abdou, dated</u> August 5, 2018.

20. See web site <u>The Significance of Leonard da Vinci's Famous "Vitruvian Man" Drawing (mymodernmet.com)</u>, for article "The Significance of Leonardo Da Vinci's Famous 'Vitruvian Man' Drawing" by <u>Kelly Richman-Abdou, dated</u> August 5, 2018.

21. See web site: Art&Object at <u>The History and Influence of Da Vinci's Vitruvian Man | Art & Object (artandobject.com)</u> for article "The History and Influence of Da Vinci's Vitruvian Man" by Peggy Carouthers.

22. See web site <u>The Significance of Leonard da Vinci's Famous "Vitruvian Man" Drawing (mymodernmet.com)</u> for article "The Significance of Leonardo Da Vinci's Famous Vitruvian Man" by Kelly Richman-Abdou, dated August 5, 2018.

23. See web site <u>Metalpoint | The Metropolitan Museum of Art (met-museum.org)</u> for article "Metalpoint".

24. See web site <u>Metalpoint | The Metropolitan Museum of Art (met-museum.org)</u> for article "Metalpoint".

25. See web site <u>The History and Influence of Da Vinci's Vitruvian Man | Art & Object (artandobject.com)</u> for article "The History and Influence of Da Vinci's Vitruvius Man".

26. See web site <u>The History and Influence of Da Vinci's Vitruvian Man | Art & Object (artandobject.com)</u> for article "The History and Influence of Da Vinci's Vitruvius Man".

27. See web site <u>Nicolaus Copernicus (Stanford Encyclopedia of Philosophy)</u> for article "Nicolaus Copernicus" revised September 13, 2019.

28. See web site <u>Nicolaus Copernicus (Stanford Encyclopedia of Philosophy)</u> for article "Nicolaus Copernicus" revised September 13, 2019.

29. See web site <u>History of Astronomy (highline.edu)</u> for the basic ideas of Copernicus' model of heliocentricity.

30. See web site <u>Nicolaus Copernicus and the Heliocentric Model - SciHi BlogSciHi Blog</u> for article "Nicolaus Copernicus and the Heliocentric Model" by Harald Sack, dated February 19, 2019.

31. See web site <u>Founding | Doctors Without Borders - USA.</u>

32. See web site <u>Independent, impartial, neutral | Doctors Without Borders - USA.</u>

33. See web site <u>Emergency response | Doctors Without Borders - USA.</u>

34. See web site <u>Types of projects | Doctors Without Borders - USA.</u>

35. See web site <u>Humanitarian AI Today on Medium. Welcome to our new Medium Publication! | by Humanitarian AI Today | Humanitarian AI Today | Medium.</u>

36. See IBM web site <u>Disaster preparedness, response and recovery (ibm.com).</u>

37. See web site <u>The technology of Doctors Without Borders | Aspiration (aspirationtech.org)</u> for article "The Technology of Borders Without Doctors", dated July 20, 2016.

38. See VSee web site <u>What is Telemedicine? All You Need To Know: Explained | VSee.</u>

39. See VSee web site <u>What is Telemedicine? All You Need To Know: Explained | VSee.</u>

40. See web site <u>What is Telemedicine? All You Need To Know: Explained | VSee.</u>

41. See web site Verdict for article "History of Smart Cities: Timeline" <u>History of smart cities: Timeline (verdict.co.uk)</u> last updated July 6, 2020.

42. See web site: Verdict <u>History of smart cities: Timeline (verdict. co.uk)</u> for article "History of Smart Cities: Timeline" last updated July 6, 2020.

43. See web site: Verdict <u>History of smart cities: Timeline (verdict. co.uk)</u> for article "History of Smart Cities: Timeline" last updated July 6, 2020.

44. See web site: Verdict <u>History of smart cities: Timeline (verdict. co.uk)</u> for article "History of Smart Cities: Timeline" last updated July 6, 2020.

45. See web site: Verdict <u>History of smart cities: Timeline (verdict. co.uk)</u> for article "History of Smart Cities: Timeline" last updated July 6, 2020.

46. See web site: McKinsey Global Institute <u>MGI-Smart-Cities-Executive-summary.pdf (mckinsey.com)</u> for article titled "Smart Cities: Digital Solutions for a More Livable Future".

47. See web site: McKinsey Global Institute <u>MGI-Smart-Cities-Executive-summary.pdf (mckinsey.com)</u> for article titled "Smart Cities: Digital Solutions for a More Livable Future".

48. See web site Santa Cruz Tech Beat <u>Geolitica: a new name, a new focus - Santa Cruz Tech Beat</u> for article "Geolitica, a new name, a new focus".

49. See web site Geolitica.com <u>Technology (geolitica.com)</u> for article "Geolitica".

50. See web site OFFICER.COM <u>Geolitica | Officer</u> for article "Geolitica".

51. See web site OFFICER.COM <u>Geolitica | Officer</u> for article "Geolitica".

52. See web site OFFICER.COM <u>Geolitica | Officer</u> for article "Geolitica".

53. See web site OFFICER.COM <u>Geolitica | Officer</u> for article "Geolitica".

54. See web site OFFICER.COM <u>Geolitica | Officer</u> for article "Geolitica".

55. See web site <u>Public Safety (geolitica.com)</u> for article "Data-Driven Community Policing".

56. See web site <u>Public Safety (geolitica.com)</u> for article "Data-Driven Community Policing".
57. See web site <u>MGI-Smart-Cities-Executive-summary.pdf (mckinsey.com)</u> for article "Smart Cities: Digital Living for a More Livable Future".

NINE

Conclusion

Homo Sapiens
is on the cusp of generating
a new version of humanity

If Neanderthals engaged Homo Sapiens in a fist fight, the Neanderthals would win, because they were physically stronger. In a cognitive competition between the two, Homo Sapiens would win, because they have stronger cognitive abilities. In the Neanderthal world, life played out in an arena of physical competition. In the Homo Sapiens world, life plays out in an arena of cognitive competition. Neanderthals became extinct. Cognitive abilities enabled Homo Sapiens to survive and thrive.

It is now the 21st century, where a competitive situation appears to be developing. Demographer, David Goodhart, informs us that the populations of Western countries are cleaving into groups of people he calls the Somewheres and the Anywheres. See Chapter 3: "David Goodhart's Trend in Demography" for a detailed description. Do the Somewheres and the Anywheres teach us anything about the future of Homo Sapiens? If there were a competition between the Somewheres and the Anywheres, who would win? Well, there were competitions in 2016. There were political competitions in the form of elections of leaders in several countries including the United Kingdom (UK), Turkey, Brazil and the United States of America (USA). The populist-leaning Somewheres won several elections, because they outnumbered the Anywheres eligible to vote in elections. The volumes of people in political parties are enough to determine election outcomes, but to what extent do elections influence the future of Homo Sapiens?

What if there were a different competition, say, about ability to live in a global culture? The Somewheres would be constrained by their geographic rootedness, and the Anywheres would win because their geographic mobility predisposes them to living in a global culture. If there were a competition about technological acumen, who would win? The Anywheres would win because their participation in a digital culture predisposes them to stronger technological skills. The Anywheres are more comfortable with globalization and technology than the Somewheres. Goodhart offers research to support the notion that the populations of Western countries are cleaving into Somewheres and Anywheres. He interprets the research to mean that the future belongs to the Somewheres, because they outnumber the Anywheres. I would point out that the Neanderthals once outnumbered Homo Sapiens. While I agree with the observation about a cleavage of populations, I disagree with Goodhart's interpretation that the future belongs to the Somewheres. I believe the future belongs to the Anywheres because they are more comfortable with globalization and technology, which I see as the drivers of the future.

Homo Sapiens differentiated itself from all the other human species, based on its cognitive abilities. In the exercise of cognitive abilities, intelligence and consciousness used to be intertwined. Cognitive abilities enable humans to acquire knowledge, turn knowledge into skills, then offload knowledge and skills onto technology. Human intelligence is being translated into products and services based on artificial intelligence. As humans offload intelligence unto tools and machines, it becomes apparent that intelligence is a transferable commodity. Human intelligence is being commoditized into products of Artificial Intelligence.

Israeli historian, Yuval Noah Harari, points out that Homo Sapiens is de-coupling intelligence from consciousness.[1] That is as significant a change as the gain in cognitive abilities that enabled Homo Sapiens to outlive Neanderthals. Cognitive abilities rely on both intelligence and consciousness. As human intelligence is being commoditized into products of Artificial Intelligence, it frees up mental resources for humans to acquire new abilities. With increasing amounts of human intelligence being relegated to technology, there appears to be an opportunity opening up for an exploration of consciousness. The empirical methods that support cognitive abilities generally have been especially influential in the development of intelligence which can be demonstrated in humanity's external world, but empirical methods are limited in enabling humans to understand consciousness which dwells in humanity's internal world.

The cognitive abilities that enable Homo Sapiens to survive are being differentiated into intelligence and consciousness. Homo Sapiens has mastered the type of intelligence that can be digitized, that is, the type that pertains to humanity's external world. Natural sciences, for example, physics and chemistry. Homo Sapiens has made less progress in the type of intelligence that pertains to humanity's internal world. Social sciences, for example, psychology and sociology. With mental resources being freed due to human intelligence being offloaded to technology, perhaps there is an opportunity to apply those free resources to exploration of consciousness.

The manner in which Homo Sapiens acquires knowledge is changing from reliance on empirical methods to an openness to discovery by algorithms. Intelligence about humanity's internal world is not directly accessible through empirical methods, but some of it is accessible through Machine Learning algorithms. Here are three examples of algorithms that give us a view into the internal world of humans. These algorithms are not driven by causality and they do not depend on empirical methods. They give us a view into humanity's internal world by discerning patterns in data collected from humans. Here are three examples of Machine Learning algorithms being used to discover patterns about the internal world of humans:

1. Google Flu Trends[2] is an algorithm used to predict the onset of influenza by country and volume of people affected. The algorithm was generated to detect influenza outbreaks early and make it possible to reduce the sickness and deaths in impacted communities. Google Flu Trends analyzed queries that people submitted to search engines about topics that mention keywords related to influenza. The queries are uncensored, unfiltered information about people's search for information related to their feelings about their wellbeing. This algorithm did not use any medical knowledge or any human medical records to predict the outbreaks. It used a search engine to find key words related to influenza in queries submitted by people who were wondering about their health. That is a novel approach; it is not the empirical way that researchers detect outbreaks of influenza. Using

an algorithm to detect patterns in searches, Google Flu Trends was able to predict the onset of influenza two weeks earlier than the Center for Disease Control (CDC), which uses empirical methods. Google Flu Trends used the text that people submit in online searches to discover a pattern in the internal world of humans, as it relates to feelings about possibly being sick with influenza.

2. Ellie[3] is an algorithm that is being tested as a virtual therapist identifying Post-Traumatic Stress Disorder (PTSD) in American military personnel who worked in Afghanistan. Ellie is not a substitute for psychologists; "she" is a screening tool. She engages patients in conversations about health issues. She is equipped with sensors that track conversation, tone of voice, changes in posture and movement in facial expressions of patients. She has access to available research on PTSD. Ellie combines these factors and analyzes them to produce an assessment in terms of probability that a patient is experiencing PTSD. Participants reported that the symptoms they present to Ellie are less censored than what they would present to human psychologists. With Ellie, patients do not have to be concerned about human psychologists being judgmental about sensitive personal matters. While human psychologists do generally observe the same factors that Ellie takes into consideration, they are not able to process the large volume of data as fast as Ellie does. An algorithm that has no consciousness, Ellie is able use patterns in conversation, voice, posture and facial expressions to discern what is in people's internal world regarding their vulnerability to PTSD.

3. Cambridge Analytica[4] was a firm that used Machine Learning in 2016 to derive psychometrics for targeting voters who are open to suggestion. Cambridge Analytica obtained Facebook data and used it to construct millions of psychographic profiles about voters in multiple countries, including the UK and USA. Cambridge Analytica realized that Facebook "likes" could be used to identify private traits with high accuracy: for example, religious affiliations, and political affiliations. The data had been leaked from academic researchers to

Cambridge Analytica, which used the Facebook psychometric data to construct millions of psychographic profiles, which it then used to hyper-target voters with custom-made campaign advertisements to favor specific candidates in UK and USA during 2016 elections. Cambridge Analytica targeted voters whose psychographic profiles indicated they were persuadable and open to suggestion. After investigations conducted by UK Parliament and US Congress, Cambridge Analytica ceased operations in 2018. Cambridge Analytica accomplished what it set out to do in developing psychographic profiles to target voters who are open to suggestion. It ceased operations because of questionable ethics, for example, permission had not been sought to use voters' data. Ethics aside, Cambridge Analytica used algorithms to gather psychometric data and generate psychological profiles about voters' internal world as it relates to political leanings and suggestibility.

Google Flu Trends, Ellie and Cambridge Analytica all used algorithms that detect patterns in data to obtain selected views into the internal world of segments of populations. Google Flu Trends used patterns in inquiries submitted via search engines to gauge the outbreaks of influenza and geographic locations of the outbreaks. Based on searches, Google Flu Trends obtained a view into the internal world of how people were feeling about their physical wellbeing. Ellie uses patterns in conversations and bodily postures of Afghan veterans to obtain a view into the internal world of their mental wellbeing. Cambridge Analytica used patterns of social media "likes" to profile voters and obtain a view into their internal world of receptivity to suggestions. The algorithms discerned patterns in data collected from humans to obtain targeted views into the internal world of humans.

Empirical methods have served us well in the past and will continue to be a valuable way of acquiring explicit knowledge in the future. Empiricism gave us competences in understanding our external world. It provided the grounding for our understanding of the physical aspects of the natural world. Then, we turned our attention to ourselves. We focused empirical methods on the physical structure of the human body, the proportions of the anatomy, the vertebrae in the skeleton, the physiology of organs, circulation of blood, and respiratory system, and genetic makeup. Still later, we turned our attention to the human mind. In the 21st century, various disciplines are bringing our focus of attention into our internal world. In earlier chapters, I draw on a variety of sources of knowledge acquisition to highlight paradigm shifts in various disciplines. The path of pattern discernment presents the opportunity for us to have algorithms augment our exploration of our internal world ... by discovering patterns about aspects of ourselves for which we might not ordinarily pose questions.

Historically, it has been easier to research humanity's external world than to research the internal world of humanity's mental activities. Humanity's internal world is not directly accessible. Researchers can ask people about their mental activity, but that assumes that people reflect on their thoughts and emotions sufficiently to be able to articulate them. That approach also assumes that people are willing to be candid about sharing their cognitive biases and unflattering emotions with researchers. At the beginning of the 20th century, psychology began to emerge as a formal academic discipline. At first the emphasis was on the classification of mental disorders and development of techniques to help people recover from mental disorders. Psychology also included the study of human development in the range of normal growth and maturation.

In the 21st century, algorithms are able to draw on data gathered from people's actual behaviors as a basis for making inferences and discerning patterns. This does not mean that empirical methods are being discarded, or that cause and effect are no longer valued. They have their value in areas where the acquisition of knowledge is guided by intentionality and rationality. Algorithmic pattern discernment has created new opportunities in areas where humans do not set out intentionally to find answers to rationally defined problems, but are searching for implicit knowledge for which we might not be able to pose questions. Pattern discernment is concerned with the use of algorithms to discover what we do not know, and what we are not aware that we could ask. Patterns include personal profiles, cognitive biases and personal habits obtained from mining data gathered from digital recordings of human activities. Some algorithms acquire explicit knowledge about humanity's external world by executing explicit instructions written by humans. Other algorithms acquire implicit knowledge about humanity's internal world by discerning patterns in large volumes of data collected from public and private sources of human activities.

New Version of Humanity with Each Revolution

Homo Sapiens is a species that emerged about 200,000 years ago in Africa and has migrated to countries all over the world. The species learned to live in geographically settled populations that produce food by domesticating animals and growing agricultural products. Over time, Homo Sapiens developed capabilities that launched the species from the Agricultural Revolution to the Scientific Revolution, to the Industrial Revolution, and to the Digital Revolution. Homo Sapiens thrived because the species developed cognitive skills that enable successful competition with other homo species. Homo Sapiens' cognitive abilities also facilitate the establishment of social structures for sustaining cultural norms in areas including politics, education, systems of justice, care for the sick, religions, as well as industrial, commercial and technological practices. Homo Sapiens takes pride in its intelligence, which is seen as the attribute that makes it superior to other species.

Over the course of revolutions, Homo Sapiens has mastered a vast array of cognitive skills, offloaded them to technology, then moved on to the next revolution. With each revolution, Homo Sapiens became more sophisticated at acquiring new knowledge and developing new skills. In the Agricultural Revolution, Homo Sapiens moved away from the nomadic lifestyle and developed skills related to establishing settled communities for domesticating animals, as well as planting and harvesting edible crops.[5] During the Scientific Revolution, Homo Sapiens reduced reliance on guidance from religious leaders and became more self-reliant in the establishment of scientific methods, which enabled an orientation to a heliocentric universe, led to the expansion of mental capabilities in rationality, and formalized education in institutions of learning.[6] As a result of the Industrial Revolution, Homo Sapiens left behind manual processes in favor of marshalling different forms of energy available to do work in transportation and manufacturing.[7] This involved developing the mental skills to render explicit knowledge into machine-performable tasks. In this revolution, Homo Sapiens used knowledge of the external world to tame nature. Homo Sapiens also developed a more sophisticated sense of autonomy while creating automobiles for transportation on self-determined journeys away from local communities. The Digital Revolution brought social connectivity via the Internet, Artificial Intelligence, a global economy, and knowledge-sharing among nations around the world.[8] These revolutions show a general historical trend. What points to an upcoming disruption in the current Digital Revolution is the increasing number of disciplines that are experiencing bifurcations.

Bifurcations in Multiple Disciplines

Long ago, Homo Sapiens differentiated itself from all the other human species, based on its cognitive ability. While other species became extinct, Homo Sapiens steadily increased its cognitive ability from one revolution to the next. Now in the 21st century, Homo Sapiens' cognitive ability is showing signs of a bifurcation. Yuval Noah Harari points out that cognitive ability is bifurcating, or splitting, into two branches: intelligence and consciousness. A bifurcation is an occurrence in a dynamical system, such as an individual, a discipline in natural or social sciences, or a human culture. A bifurcation is an abrupt change in the behavior of a dynamical system, due to one or more parameters in the system reaching a threshold value.[9] There are local bifurcations and global bifurcations. A local bifurcation is a change in one parameter that disrupts the equilibrium of a dynamical system. A global bifurcation is a collection of changes in multiple parameters that together disturb the equilibrium of a community of interacting systems.

A bifurcation is a precursor to qualitative change in a dynamical system. In each chapter of this book, I describe a paradigm shift in a discipline that is either in natural sciences or social sciences. Each paradigm shift involves a bifurcation that disturbs the equilibrium of that discipline. These are local bifurcations. Taken together, the local bifurcations constitute a global bifurcation that has the potential to disrupt the equilibrium of Western culture. The disciplines described in this book are examples from a larger collection of disciplines that make up a global bifurcation. This book argues that a growing global bifurcation puts Homo Sapiens on the cusp of a major disruption, which I believe will generate a new version of humanity.

Here are the local bifurcations in each discipline described in this book:

- **Technology:** The research and consulting firm, Gartner Incorporated, discerns a paradigm shift from technology-literate people, to people-literate technology.[10] This shift is moving from a situation where people learn about technology in the external world, to a situation where technology is learning about people. That includes aspects of humanity's internal world. The local bifurcation in this discipline is in the parameter of literacy. Literacy used to be attributed to humans. Now, literacy is bifurcating, that is, literacy is being split into human literacy and technological literacy. Humans and technology are both capable of learning, however, they learn in different ways.
- **Psychology:** Psychologist Wolfgang Giegerich points to a paradigm shift from a focus on the psychology of individuals (individuation process), to a focus on the psychology of human culture (interiorization process).[11] This is a shift from the semantical level of psychology, where humans expand consciousness by incrementally, to the syntactical level of psychology, where

humans restructure consciousness enabling an internal world that supports higher, more complex capabilities. The local bifurcation in this discipline is in agency, which is splitting into agency of individuals and agency of human culture.

- **Demography:** Demographer David Goodhart reports on a paradigm shift that takes the form of a cleavage of Western populations into groups of "Somewheres" and "Anywheres".[12] This shift is from a traditional grouping of people whose identity is inherited from the external world of family and community, to a contemporary group of people whose identity is created from their internal world of self-selected achievements. The local bifurcation in this discipline is identity, which is splitting into inherited identity and achieved identity.

- **Business Operations:** The research and consulting firm, Gartner Incorporated, discerns a paradigm shift from a business model of (Somewhere) Operations, to a business model of Anywhere Operations.[13] This shift is from a traditional model of business operations whose boundaries are fixed by time and geography, to a contemporary model of business operations, where employees, customers and vendors share an internal mindset that blurs boundaries of time and geography. The local bifurcation in this discipline is business model. The parameter of business model is bifurcating into physical operations and digital operations.

- **Project Management:** The Project Management Institute (PMI) identifies a paradigm shift from a practice of managing projects in traditional enterprises to managing projects in gymnastic entrprises.[14] The shift is from supporting traditional enterprises that have a lesser focus on building a digital culture, to gymnastic enterprises that have a greater focus on building a digital culture. The local bifurcation in this discipline occurs in the practice of project management, which is splitting into data management and strategy management.

- **Genetic Engineering:** Nobel Prize winning chemist, Jennifer Doudna, writes about a paradigm shift from a practice of gene therapy that treats genetic disease in individuals, to a practice of germline enhancement that changes heritable characteristics in future generations.[15] This is a shift from recovery to prevention. The local bifurcation in this discipline is treatment for genetic disease, which is splitting into healing humans and designing humans.

- **Reputation Management:** Privacy law expert, Daniel Solove, sees a paradigm shift from a custom of individuals managing their own reputations by selecting from their achievements in the external world, to a custom of algorithms managing people's reputations by gathering information from social media and other digital sources where private information is not protected.[16] This is a shift from a situation where individuals curate the reputation they choose to present to the community, to a situation where society curates people's reputations. The local bifurcation in this discipline is a splitting of the control of reputation management. Reputation management is bifurcating into personal control of reputation management and societal control of reputation management.

- **Biocentrism:** A research scientist in biology, Robert Lanza, discerns a paradigm shift from a focus on the consciousness of individuals who are detached observers of reality, to a focus on the consciousness of collective groups that are active co-creators of reality.[17] Biocentrism makes a distinction between the consciousness of individuals and the consciousness of collective groups, in terms of how they relate to physical reality. The local bifurcation in this discipline is the attribute of consciousness. The parameter of consciousness is bifurcating into consciousness of individuals and consciousness of collective groups.

Taken together, the local bifurcations in these disciplines are indicative of a growing global paradigm shift that places Homo Sapiens on the edge of a major disruption of the Digital Revolution. I propose that the disruption will generate a new revolution out of which will emerge a new version of humanity.

NOTES:

1. See "Homo Deus: A Brief History of Tomorrow" by Yuval Noah Harari, HarperCollins, 2017, p 314.
2. See Google's web site: Google Flu Trends Estimates - Google Public Data Explorer for the article "Google Flu Trends Estimates".
3. See NBC News web site: How Virtual Therapy Could Help the Military Fight PTSD (nbcnews.com) for the article "How Virtual Therapy Could Help the Military Fight PTSD".
4. See web site: Frontiers in Psychology Frontiers | Machine Learning in Psychometrics and Psychological Research | Psychology (frontiersin.org), Article "Machine Learning in Psychometrics and Psychological Research" by Graziella Orru, Merylin Monaro, Ciro Conversano, Angelo Gemignani and Guiseppe Sartori.
5. See web site: ScienceDirect, Agricultural Revolution - an overview | ScienceDirect Topics, Article "Agricultural Revolution".
6. See web site: Britannica, Scientific Revolution | Definition, History, Scientists, Inventions, & Facts | Britannica, Article "Scientific Revolution".
7. See web site: History, Industrial Revolution: Definitions, Causes & Inventions - HISTORY, Article "Industrial Revolution".
8. See web site: Forbes, The Fourth Revolution: The Age Of Digital Enlightenment (forbes.com), Article "The Fourth Revolution: The Age of Digital Enlightenment".
9. See "Dynamic Patterns: The Self-Organization of Brain and Behavior" by J. A. Scott Kelso, The MIT Press, 1995, Bifurcation pp 85 – 87.
10. See Gartner's web site for the paradigm shift from Technology-Literate People to People-Literate Technology: https://www.gartner.com/en/newsroom/press-releases/2019-10-21-gartner-identifies-the-top-10-strategic-technology-trends-for-2020.
11. See Giegerich's observation of a paradigm shift in psychology in "The Soul Always Thinks" pp 329 – 330.
12. See the book "The Road to Somewhere: The Populist Revolt and the Future of Politics" by demographer David Goodhart, pp 3 - 13.

13. See Gartner web site: https://www.gartner.com/smarterwithgartner/gartner-top-strategic-technology-trends-for-2021/.

14. See the Project.net web site: An Introduction to Algorithms for Solving Schedule-Related Problems | Open Source Project Management Software - Project.net.

15. See the book "*A Crack In Creation: Gene Editing and the Unthinkable Power to Control Evolution*" co-authored by Jennifer A. Doudna and Samuel H. Sternberg, pp 213 - 240.

16. See the book "*The Future of Reputation: Gossip, Rumor, and privacy on the Internet*" by international expert in privacy law, Daniel J. Solove, pp 362 - 394.

17. See the book "*The Grand Biocentric Design: How Life Creates Reality*" by Robert Lanza, pp 187 - 195.

APPENDIX A

Consciousness

To explain consciousness, I take an excerpt from "A Critical Dictionary of Jungian Analysis" by psychologists Andrew Samuels, Bani Shorter and Fred Plaut, pp 36 – 37.

> "At various times, (Carl) Jung equated consciousness with awareness, intuition and apperception, stressing the function of reflection in its achievement. Attainment of consciousness would appear to be the result of recognition, reflection upon and retention of psychic experience, enabling the individual to combine it with what he has learned, to feel its relevance emotionally, and to sense its meaning for his life.

> …..

> Reaching the conclusion that the most individual thing about man was his consciousness, and based on the supposition that individuation is a psychic necessity, Jung's psychology became equated with increased consciousness, and in analysis the assumption was that consciousness would shift from ego centeredness toward a point of view more consistent with the totality of the personality."

APPENDIX B

Individuation

This explanation of individuation is taken from "*A Critical Dictionary of Jungian Analysis*" by psychologists Andrew Samuels, Bani Shorter and Fred Plaut, pp 76 - 79.

> "A person's becoming himself, whole, indivisible and distinct from other people or collective psychology (though also in relation to these).

This is the key concept in (Carl) Jung's contribution to the theories of personality development. As such, it is inextricable interwoven with others, particularly SELF, EGO, and ARCHETYPE as well as with the synthesis of CONSCIOUSNESS and UNCONSCIOUS elements. A simplified way of expressing the relationship of the most important concepts involved would be: ego is to INTEGRATION (socially seen as ADAPTATION) what the self is for individuation (self-experience and realization). While consciousness is increased by the analysis of defenses (e.g., PROJECTION of the SHADOW), the process of individuation is a CIRCUMAMBULATION of the self as the centre of the personality which thereby becomes unified. In other words, the person becomes conscious in what respects he or she is both a unique human being and at the same time, no more than a common man or woman.

Because of the inherent paradox, definitions abound, both throughout Jung's work as well as that of the 'post-Jungians' (Samuels, 1985a) The term 'individuation' was taken up by Jung via the philosopher Schopenhauer but dates back to Gerard Dorn, a sixteenth-century alchemist. Both speak of the principium individuationis. Jung applied the principle to psychology.

…..

The attributes emphasized are:

(T)he goal of the process is the development of the personality

(I)t presupposes and includes COLLECTIVE relationships, i.e., it does not occur in a state of isolation

(I)ndividuation involves a degree of opposition to social norms which have no absolute validity.

…..

Individuation is no more than a potential goal, the idealization of which is easier than the realization.

…..

(S)ymbols of the self (e.g., mandalas and dreams) … occur wherever the process of individuation 'becomes the object of conscious scrutiny', or where, … the collective unconscious peoples the conscious mind with archetypal figures' (CW 16, para. 474)."

APPENDIX C

Giegerich: "Soul", "Psyche" and "Autonomous Mind"

In describing the interiorization process, Giegerich often makes reference to the "soul" which is a word he uses interchangeably with psyche, objective psyche, autonomous mind, among others. The following are examples of the ways in which he uses of the expressions soul, psyche and autonomous mind.

> "The <u>soul</u> is fundamentally a linguistic reality. As self-representation, the soul in all its countless and most variegated forms of expression has no other topic but itself, its own truths. As self-relation it is also self-referential."

See "What Is Soul?" by Wolfgang Giegerich, p 44.

> " '<u>Psyche</u>' integrates the soul into the scientific notion of the natural world at large."

See "What Is Soul?" by Wolfgang Giegerich, p 16.

"The <u>psyche</u> engages itself in a dialogue; the topic of the dialogue is itself, its own inner nature."

See "What Is Soul?" by Wolfgang Giegerich, p 44.

"The soul's logical life or what Jung called the <u>autonomous mind</u>, … is essentially productive, in fact ongoing production (without a producer). It is self-expression, self-representation, self-portrayal. In this sense it is, both in a literal and in a general sense, speaking as such (a speaking, however, without a speaker)."

See "What Is Soul?" by Wolfgang Giegerich, p 44.

I underlined the expressions soul, psyche and autonomous mind, to make them obvious to readers.

APPENDIX D

Giegerich: Soul Movements

This Appendix contains my interpretation of three of Giegerich's Soul Movements that I consider relevant to the emergence of technology as a phenomenon in the external world. They are "internal contradiction", "procreation" and "opus of nature conquering nature".

Giegerich's Soul Movement: Internal Contradiction

One soul movement proceeds on the basis of internal contradictions between identity and difference within the soul. See "*The Soul Always Thinks*" pp 308 – 311 for reference to internal contradiction.

The movement is made up of serial opposites of contradiction. The internal contradiction is called the dialectical movement. This movement has its non-physical existence only in the realm of thought. It is not a natural movement of any physical substance in time. It is the movement of thought itself. The movement is self-enclosed and abstract. Internal contradiction is the dialectical movement of thought that results in the creation of new thought.

I see the dialectical movement reflected in Gartner's paradigm shift from a model of technology-literate people to a model of people-literate technology. Gartner's paradigm shift is not physical. It is about a mental modelling of behavior in the external world. The mental model begins with people becoming literate enough about technology to build artificially intelligent systems. The internal contradictions are about intelligence. The initial outlook is that intelligence is an attribute of humans. People use their intelligence to build technology. Gradually, through internal contradictions, intelligence moves back and forth between being an attribute of people and being an attribute of technology. Intelligence becomes an attribute of technology, in the form of Artificial Intelligence. Through internal contradictions, the mental model shifts to technology becoming intelligent enough to assemble histories and build profiles of people.

Giegerich's Soul Movement: Procreation

Another soul movement consists of the soul's need to break its own self-containment, within its circular movement and open itself to release itself from the sphere of thought, and enter into the real world of nature. See "*The Soul Always Thinks*" pp 311 – 314 for reference to procreation.

The pure thought of the soul wants to immerse itself in the sensory world. I see procreation as the productiveness of the soul's movements in the creation of new technological concepts before they are released into the world. When released into the world, it appears as the spontaneous, self-manifestation of nature, in the form of phenomena and epiphanies of their own accord happening to people and showing the deeper truth of the world. The soul's procreation is not a material-factual production. In the procreation movement, humans beget Machine Learning which begets self-learning algorithms. I regard Machine Learning as an example of an epiphany in technology. The idea that Machine Learning enables algorithms that can learn on their own is innovative, but the amazing revelation is that algorithms can now acquire literacy about the people who created Machine Learning, without the awareness or permission of people involved.

Giegerich's Soul Movement: Opus of Nature Conquering Nature

In this type of soul movement, the soul focuses on one particular work, called an opus, and sustains the effort until there is a shift from the current status of consciousness to a higher status of consciousness. See *"The Soul Always Thinks"* pp 316 – 323 for reference to opus of nature conquering nature.

This involves a destructive, negating, sublating work imposed upon the current status of consciousness, as if from outside the containment of the soul, with a view to changing consciousness to a new level of truth, that is, a new level of content. The sense of the current consciousness is sublated and refined, and a new consciousness begins to form. This type of soul movement is propelled forward to ever new stages of itself, ever new statuses of consciousness.

Other types of soul movement address the natural world. Opus of nature conquering nature is the movement from one status of consciousness to a higher status where a new level of truth results in the creation of a new outlook on the world. The opus addresses consciousness itself, consciousness' distillation and further development. In my opinion, there is an opus, a work of Great Technology in Gartner's paradigm shift. The shift is that there will be a trend from a model of technology-literate people to a model of people-literate technology. That shift is already becoming evident in the fact that Machine Learning enables algorithms to become literate about people.

APPENDIX E

Giegerich: Dialectical Movement

The dialectical movement in the psyche results in a history of the development of consciousness. This is how Giegerich describes the development of consciousness.

> "Development here means consciousness is pushing off from its initial stage, negating and sublating the latter and ipso facto reaching a new stage, from which the same process can begin again. ... In each of the new stages reached the psyche realizes and fulfills one potential of itself and brings out into the open one more aspect of its own nature."

See *"The Soul Always Thinks"* p 349.

I underline phrases in Giegerich's wording to show their relationship to the dialectical movement:

- "Pushing off from its initial stage" corresponds to movement away from the initial outlook, or existing stage of consciousness.
- "Negating and sublating the latter" correspond to opposition to the existing stage of consciousness. Repeated negation of the consciousness that exists at the time constitutes a series of contradiction of opposites, that produce versions of consciousness.

- "Reaching a new stage" corresponds to the achievement of a new outlook, that involves a higher stage of consciousness.

Here is Giegerich's further explanation of the dialectical movement:

The dialectical movement is about "... the self-exposure to and the self-immersion in the process of the conscious standpoint's going under, (to) its self-distillation. Inwardization and recursive progression instead of engulfment. ... (I)t is the self-distillation, self-liquefaction of all semantics of the metaphysical tradition, into syntactical or logical form. ... We could say: it is the assimilation or integration or inwardization of the content of metaphysics into the form of consciousness. ...Sublation is not a goal toward which one could be on the way. The 'not' is neither a reminder nor 'a momentary hiatus in the dialectic'. ...It is not something separate or in between two statuses or phenomena, not what happens to, is done to, or is inflicted upon a given stance. It is the realization that the matter has all along not been what it had seemed to be. And this is at once the (first immediacy of the) recognition of the new form of the matter."

See "What Is Soul?" pp 298 - 299.

APPENDIX F

Autopoiesis

In the 1970s, two Chilean researchers co-authored a book titled "The Tree of Knowledge" in which they introduced the term "autopoiesis". See "*The Tree of Knowledge*" co-authored by Humberto Maturana and Francisco Varela, Shambala 1987.

Biologist Humberto Maturana and cognitive scientist Francisco Varela used the term "autopoiesis" to refer to early biological life-forms that are capable of reproducing and maintaining themselves. They defined autopoiesis as the mechanism that characterizes living beings as autonomous systems. They further explained that human beings are included among the biological creatures characterized by their autopoietic organization, that is, they are continually self-producing.

This is how Maturana and Varela explain the autopoietic mode of organization:

> "What is distinctive about (living beings) … is that their organization is such that their only product is themselves, with no separation between producer and product. The being and doing of an autopoietic entity are inseparable, and this is their specific mode of organization."

See Maturana and Varela, pp 48 – 49.

As further explanation, Maturana and Varela describe autopoiesis in terms of structural coupling:

> " ... the interaction between (an autopoietic entity) and environment will consist of reciprocal perturbations. In these interactions, the structure of the environment only triggers structural change in the autopoietic entity (it does not specify or direct them), and vice versa for the environment. The result will be a history of mutual congruent structural changes as long as the autopoietic entity and its containing environment do not disintegrate: there will be a structural coupling." See Maturana and Varela, p 75.

An autopoietic system is considered operationally closed, because it contains what it needs to maintain itself as a whole, without requiring resources from outside itself. An autopoietic system is structurally coupled with the environment in which it is embedded. The structured coupling is a history of interactions between the autopoietic system and its environment. Although the autopoietic system and the environment function independently of each other, the interaction between them is affected by a continuous dynamic that is regarded as a type of knowledge acquisition.

Glossary

WORD	MEANING
Algorithm	A self-contained set of software procedures for processing data.
Analytical Psychology	A school of psychology that was developed by Carl Jung. It studies both the conscious and unconscious aspects of the human psyche, and promotes a quest for individuation.
Apps	Software applications that are designed to be used on mobile devices, such as smartphones.
Archetype	An inherited component of the human psyche that has the capacity to influence behaviors. Archetypes are structures of the psyche that constitute pre-dispositions for people to think, feel, perceive and act in specific ways. Archetypes populate the unconscious part of the psyche.
Artificial Intelligence	A branch of computer science in which computer systems perform tasks that would otherwise be performed by humans.

WORD	MEANING
Bifurcation	A split in a parameter of a self-organizing system, such as human biology. A characteristic of self-organizing systems is that when a system changes enough, it brings about turmoil. When collective changes in a self-organizing system reach a critical point, they create turmoil in the system and bring about bifurcations, which create the potential for growth.
Chaos Theory	The interdisciplinary science which states that although complex systems may appear to be random and unpredictable, there are underpinning patterns. When there is chaos in a system, it creates an opportunity for the emergence of novel phenomena.
Cognitive Psychology	A school of psychology founded by Ulric Neisser. It focuses on the conscious aspect of the human psyche often by noting sensory input, applying stimuli and analyzing subsequent behavioral changes.
Collective Unconscious	That part of the psyche that contains experiences shared by all humanity.
Complex	A word that Carl Jung used to mean a core pattern of emotion arranged around a common theme and located in the personal unconscious part of the human psyche.
Data Mining	A branch of computer science that detects patterns in large datasets, in order to generate new knowledge.
Ego	That component of the human psyche which mediates interaction between consciousness and unconsciousness.

WORD	MEANING
Individuation	A goal-seeking effort of psychological development where individuals develop their uniqueness by differentiating their minds from the unconsciousness of their communities.
Interiorization	A focus of attention on cultural knowing for engaging in an intellectual discipline of interpreting phenomena that emerge in the world.
Machine Learning	A branch of computer science where algorithms learn in an iterative manner by discerning patterns in large volumes of data that are being continuously updated.
Persona	The social face that an individual shows to others to hide the true nature of the individual.
Personal Unconscious	That part of the psyche that contains repressed or forgotten experiences of an individual.
Projection	The involuntary casting of unconscious material onto an entity in the external world.
Psyche	The collection of processes and content that pertain to both consciousness and unconsciousness.

Bibliography

Davies, Kevin
  "*Editing Humanity: the CRISPR Revolution and the New Era of Genome Editing*"
  Penguin Books, 2020

Doudna, Jennifer A. & Sternberg, Samuel H.
  "*A Crack In Creation: Gene Editing and the Unthinkable Power to Control Evolution*"
  First Mariner Books, 2018

Giegerich, Wolfgang
  "*Technology and the Soul: From the Nuclear Bomb to the World Wide Web*"
  The Collected English Papers of Wolfgang Giegerich, Volume 2
  Routledge, 2007

Giegerich, Wolfgang
  "*The Soul Always Thinks*"
  The Collected English Papers of Wolfgang Giegerich, Volume 4
  Routledge, 2020

Giegerich, Wolfgang
  "*What is Soul?*"
  Spring Journal Books, 2012

Gleick, James
"*CHAOS: Making a New Science*"
Penguin Books, 1988

Goodhart, David
"*The Road to Somewhere: The Populist Revolt and the Future of Politics*"
Hurst Publishers, 2017

Harari, Yuval Noah
"*Homo Deus: A Brief History of Tomorrow*"
HarperCollins Publishers, 2017

Harari, Yuval Noah
"*Sapiens: A Brief History of Humankind*"
HarperCollins Publishers, 2015

Isaacson, Walter
"*The Code Breaker: Jennifer Doudna, Gene Editing, and the Future of the
Human Race*"
Simon & Schuster, 2021

Jung, Carl Gustav
"*The Structure of the Unconscious*"
Collected Works, Volume 7
Princeton University Press, 1972

Lanza, Robert & Pavsic, Matej & Berman, Bob
"*A Grand Biocentric Design: How Life Creates Reality*"
BenBella Books, 2020

Lanza, Robert (editor)
"*ONE WORLD: The Health and Survival of the Human Species in the
21st Century*"
Health Press, 1996

Markowits, Daniel
  "*The Meritocracy Trap*"
  Penguin Press, 2019

Maturana, Humberto & Varela, Francisco
  "*The Tree of Knowledge*"
  Shambala 1987

Robertson, Robin & Combs, Allan (editors)
  "*Chaos Theory in Psychology and the Life Sciences*"
  Lawrence Erlbaum Associates, 1995

Samuels, Andrew & Shorter, Bani Shorter & Plaut, Fred
  "*A Critical Dictionary of Jungian Analysis*"
  Routledge & Kegan Paul Ltd., 1993

Solove, Daniel J.
  "*The Future of Reputation: Gossip, Rumor, and Privacy on the Internet*"
  Yale University Press, 2008

Stein, Murray
  "*The Principle of Individuation: Toward the Development of Human
     Consciousness*"
  Chiron Publications, 2018

Wylie, Christopher
  "*Mindf*ck: Cambridge Analytica And The Plot To Break America*"
  Random House, 2019

Index

T

U

V

W